# EXPERIMENTAL MANUAL IN MEDICAL BIOCHEMISTRY

FIRST EDITION

Chief Editors
YU Hong and Huang Xinxiang

Subeditors
HE Chunyan, GUAN Yaqun, ZHANG Baifang, GE Yinlin

WUHAN UNIVERSITY PRESS
武汉大学出版社

图书在版编目(CIP)数据

EXPERIMENTAL MANUAL IN MEDICAL BIOCHEMISTRY/YU Hong and Huang Xinxiang Chief Editors. —武汉:武汉大学出版社,2008.9
ISBN 978-7-307-06534-5

Ⅰ.E… Ⅱ.①Y… ②H… Ⅲ.生物化学—实验—医学院校—教材—英文 Ⅳ.Q5-33

中国版本图书馆 CIP 数据核字(2008)第 143850 号

---

责任编辑:黄汉平　　责任校对:刘　欣　　版式设计:马　佳

出版发行:武汉大学出版社　(430072　武昌　珞珈山)
　　　　　(电子邮件:wdp4@whu.edu.cn　网址:www.wdp.com.cn)
印刷:湖北金海印务公司
开本:787×1092　1/16　印张:12.75　字数:283 千字　插页:1
版次:2008 年 9 月第 1 版　　2008 年 9 月第 1 次印刷
ISBN 978-7-307-06534-5/Q·90　　　定价:22.00 元

版权所有,不得翻印;凡购买我社的图书,如有缺页、倒页、脱页等质量问题,请与当地图书销售部门联系调换。

# Contributors

| | | |
|---|---|---|
| 卜友泉 | Bu Youquan | Chongqing Medical University, Chongqing |
| 蔡望伟 | Cai Wangwei | Hainan Medical College, Hainan |
| 曹 佳 | Cao Jia | College of Medicine, Wuhan University, Wuhan |
| 陈 艳 | Chen Yan | Xinjiang Medical University, Xinjiang |
| 杜 芬 | Du Fen | College of Medicine, Wuhan University, Wuhan |
| 葛银林 | Ge Yinlin | Qingdao University Medical College, Qingdao |
| 关亚群 | Guan Yaqun | Xinjiang Medical University, Xinjiang |
| 何春燕 | He Chunyan | College of Medicine, Wuhan University, Wuhan |
| 候 敢 | Hou Gan | Guangdong Medical College, Guangdong |
| 黄冬爱 | Huang Dong'ai | Hainan Medical College, Hainan |
| 黄新祥 | Huang Xinxiang | Jiangsu University, Jiangsu |
| 林春榕 | Lin Chunrong | Dali College, Dali |
| 马永平 | Ma Yongping | Chongqing Medical University, Chongqing |
| 生秀梅 | Sheng Xiumei | Jiangsu University, Jiangsu |
| 孙续国 | Sun Xuguo | Tianjin Medical University, Tianjin |
| 唐 微 | Tang Wei | Yunyang Medical College. Yunyang |
| 涂建成 | Tu Jiancheng | Wuhan University, Wuhan |
| 王继红 | Wang Jihong | Chongqing Medical University, Chongqing |
| 武军驻 | Wu Junzhu | College of Medicine, Wuhan University, Wuhan |
| 肖方祥 | Xiao Fangxiang | Medical Science College of China Three Gorges University, Yichang |
| 燕 秋 | Yan Qiu | Dalian Medical University, Dalian |
| 杨雪松 | Yang Xuesong | Dalian Medical University, Dalian |
| 杨艳燕 | Yang Yanyan | Hubei University, Hubei |
| 喻 红 | Yu Hong | College of Medicine, Wuhan University, Wuhan |

Experimental Manual in Medical Biochemistry

| 张百芳 | Zhang Baifang | College of Medicine, Wuhan University, Wuhan |
| 郑 芳 | Zheng fang | Wuhan University, Wuhan |
| 周宏博 | Zhou Hongbo | Harbin Medical University, Harbin |
| 朱名安 | Zhu Ming'an | Yunyang Medical College, Yunyang |
| 左绍远 | Zhuo Shaoyuan | Dali College, Dali |

# PREFACE

Progress in life science, including medical science, is mainly made by observation and experiment, especially biochemistry and molecular biology experiment. So, learning the basic biochemical techniques and methods is essential to medical students and the students of other related disciplines.

The first edition of this manual presents shortened versions of the basic biochemistry methods designed for use at the lab bench in a biochemistry and molecular biology laboratory. It is primarily intended for undergraduate medical students, and hopefully it will be a useful reference book for graduate students and teaching staff. The objectives for the course are for students to learn modern biochemical laboratory methodologies and techniques, develop the ability to perform experiments successfully and independently, properly interpret the data and write scientific laboratory reports. Students who learn biochemistry as a part of medical science should be able to understand the concepts and methods of not only the conventional clinical laboratory tests but also the recently developed tests. For this purpose, we have made this manual with incorporation of some clinical laboratory test kits. All of the experiments have been performed many times in our own laboratory. We hope that they will provide a reliable training of the most important and commonly used techniques of contemporary biochemistry.

We express our appreciation to our colleagues in the Department of Biochemistry and Molecular Biology, Wuhan University School of Medicine, who have contributed ideas, techniques and experiences to the manual. We are really grateful to all the experts, especially those from other universities, who made big contribution to the compiling of various chapters.

This manual is not a complete version. It should be continually modified and updated. We would be very grateful if the users of this manual could bring good feedbacks for the development of this manual.

**Yu hong and Huang xinxiang**
July, 2008

# Requirements

**Lab safety**

1. Eating, drinking, smoking or chewing gum in the laboratory is strictly prohibited.
2. Always wear lab-coat, sandals are not allowed in the laboratory.
3. Keep your work area clean. When chemicals are spilled they should be wiped as soon as possible. Be aware of objects that can burn or give electrical shocks.
4. Mouth pipetting is not allowed.
5. Do not turn on an instrument until you have read instructions and consulted the instructors for the use of equipment. If any equipment malfunction is noted, report this immediately.
6. Anyone carrying out these protocols will encounter the following hazardous materials: (1) toxic chemicals and carcinogenic or teratogenic reagents, (2) pathogens and infectious biological agents. We emphasize that users must proceed with the prudence and precaution associated with good laboratory practice. Use chemicals with high vapor pressure only in the hood. Handle and dispose of hazardous chemicals properly. Disposal containers are provided.
7. In case of an accident, notify an instructor immediately. For any chemicals splashed in the eye, hold the eye open and flush immediately with cold water using the eye wash. For chemicals spilled on the skin or splashed into the mouth, flush with large amounts of cold water. For burns, flush with cold water and contact an instructor.

**Attendance policy and lab reports**

Students are expected to attend every lab, and to arrive promptly and well prepared. A student who is absent without the prior permission of the instructor or does not have documented excuse will receive 0 points on the lab report for that experiment.

All lab reports must be done on an individual basis. You will be given instructions about the format and the information needed for each experiment. Laboratory reports must include the following: experiment title, date(s) experiment performed, experimental results (data, method of calculations, tables, graphs, figures, significance of the data), discussions.

# Content

## Chapter I  Spectrophotometry ............... 1
*He Chunyan, Wang Jihong, Du Fen, Zhu Ming'an, Tu Jiancheng*
1.1 Basic Concepts and Principle of Spectrophotometry ............... 1
1.2 Measurement of Light Absorption ............... 4
1.3 Aplications of Spectrophotometry ............... 6
EXP. 1 Quantitative Analysis of Protein ............... 8
EXP. 2 Determination of Blood Glucose ............... 18

## Chapter II  Electrophoresis ............... 23
*Huang Xinxiang, He Chunyan, Hou Gan, Sheng Xiumei*
2.1 Basic Principles of Electrophoresis ............... 23
2.2 Impact Factors of Electrophoresis ............... 24
2.3 Detection of Components after Electrophoretic Separation ............... 26
2.4 Special Electrophoretic Techniques and Applications ............... 27
EXP. 3 Separation of Serum Proteins by Cellulose Acetate Membrane Electrophoresis ............... 31
EXP. 4 Separation of Serum Lipoproteins by Agarose Gel Electrophoresis ............... 35
EXP. 5 Proteins Separation by SDS-Polyacrylamide Gel Electrophoresis ............... 38

## Chapter III  Chromatography ............... 44
*Xiao Fangxiang, Yang Xuesong, Lin Chunrong, Yang Yanyan*
3.1 Fundamental Principles of Chromatography ............... 44
3.2 Commonly Used Chromatographic Techniques ............... 46
EXP. 6 Detecting Transamination of Amino Acids by Paper Chromatography ............... 54
EXP. 7 Separation of Mixed Amino Acids by Cation Exchange Chromatography ............... 58

EXP. 8 Separation of Hemoglobin and Dinitrophenyl (DNP)-glutamate by Gel Filtration Chromatography ............ 62

## Chapter IV Enzyme Analysis ............ 65

*Huang Dong'ai, Cai Wangwei, Sun Xuguo, Zhuo Shaoyuan*

4.1 The Activity of Enzyme ............ 65
4.2 Enzyme kinetics (Factors affecting reaction velocity) ............ 67
4.3 Enzymatic analysis ............ 70
EXP. 9 Assay the Activity of Alanine Aminotransferase (ALT) in Serum (Mohun's Method) ............ 74
EXP. 10 Lactate Dehydrogenase (LDH) Analysis ............ 78
EXP. 11 Effect of Substrate Concentration on Enzyme Activity-Determining Km Value for Alkaline Phosphatase ............ 83
EXP. 12 Coupled enzymatic reaction assay (EnzymaticEndpoint Method)-Determination of Total Cholesterol, Triglycerides and Glucose in Serum ............ 86

## Chapter V Isolation and Purification of Protein ............ 94

*Yu Hong, Cao Jia, Guan Yaqun, Zhou Hongbo, Chen Yan*

5.1 Methods and Basic Principles of Protein Purification ............ 94
5.2 Procedure of Protein Preparation ............ 100
5.3 Crystallization, Concentration, Drying and Storage of Protein ............ 103
5.4 Identification and Analysis of Protein ............ 104
EXP. 13 Isolation and Identification of Serum IgG ............ 107
EXP. 14 Expression, Purification and Identification of Glutathione S-Transferase Fusion Protein ............ 114

## Chapter VI Isolation, Purification and Identification of Nucleic Acids ............ 121

*Zhang Baifang, Yan Qiu, Ma Yongping, Zheng Fang*

6.1 Isolation and Purification of Nucleic Acids ............ 121
6.2 Identification and Analysis of Nucleic Acids ............ 124
EXP. 15 Isolation of Eukaryotic Genomic DNA by Proteinase K-Phenol or NaI Method ............ 132
EXP. 16 SDS-Alkaline Lysis of Plasmid DNA ............ 138
EXP. 17 Isolation of Total RNA by TRIzol Reagent ............ 141

EXP. 18    Identification of DNA by UV-Spectrophotometry and
           Gel Electrophoresis ............................................................ 145
EXP. 19    Identification of RNA by UV-Spectrophotometry and Formaldehyde
           Denaturating Agarose Gel Electrophoresis ........................... 153
EXP. 20    Polymerase Chain Reaction (PCR) and Reverse Transcription-PCR ...... 157

## Chapter VII    Genetic Engineering .................................... 163

*Ge Yinlin, Wu Junzhu, Bu Youquan, Tang Wei*

7.1    The Process of Gene Cloning ............................................. 163
7.2    Enzymes for Gene Cloning ................................................ 165
7.3    Vectors for Gene Cloning .................................................. 166
7.4    Expression of Foreign Genes and Purification of Recombinant Proteins ...... 168
EXP. 21    DNA Cloning ................................................................. 171
EXP. 22    Expression of Exogenous Gene in *E. coli* and Purification of the
           Expressional Product ....................................................... 180

## English-Chinese Index ................................................ 190

EXP. 17. Isolation of DNA by UV-spectrophotometry and Gel Electrophoresis .................................................. 145

EXP. 18. Identification of RNA by UV-spectrophotometry and Formaldehyde Denaturation Agarose Gel Electrophoresis .................................................. 151

EXP. 20. Polymerase Chain Reaction (PCR) and Reverse Transcription-PCR .................................................. 157

## Chapter VII  Genetic Engineering .................................................. 163

7.1 The Process of Gene Cloning .................................................. 163
7.2 Enzymes for Gene Cloning .................................................. 165
7.3 Vectors for Gene Cloning .................................................. 166
7.4 Expression of Foreign Genes and Purification of Recombinant Proteins .................................................. 171
EXP. 21. DNA Cloning .................................................. 179
EXP. 22. Expression in Escherichia coli of rhEGF gene and Purification of the Biochemical Products .................................................. 182

## English-Chinese Index .................................................. 191

# Chapter I
# Spectrophotometry

Spectrometry is a qualitative and quantitative technology based on emission spectra, absorption spectra or scattering spectra of the material. Spectrometry can be classified in many different types depended on the form of spectra, including atomic emission spectrometry, flame photometry and fluorescence spectrometry based on the character of emission spectra, UV and visible spectrophotometry based on the absorption spectra, atomic absorption and infrared spectroscopy based on scattering spectral character.

Spectrophotometry is based on the reflection or transmission properties of a substance. It has some advantages, such as simple, rapid, high sensitivity, accuracy, and selectivity. So, it is one of the most widely used analytical techniques in biochemistry laboratory and clinical research.

## 1.1 Basic Concepts and Principle of Spectrophotometry

Light is a kind of electromagnetic wave. The light can be seen by our naked eyes is really a very small portion of the electromagnetic spectrum (Figure 1-1) within the wavelength range between approximately 380 nm and 760 nm. The visible light could be split into many colors by a prism. Each color is signed by the wavelength of light. Any solution that contains a subtance that absorbs the visible light will appear a color. Light with wavelengths longer than 760 nm or shorter than 380 nm is invisible. Ultraviolet is a region of the electromagnetic spectrum that has a wavelength range from 375 nm to 12.5 nm. Ultraviolet-Visible spectrophotometry using lights in the visible range and adjacent near ultraviolet (UV) range (200~375 nm) is discussed in this chapter.

### 1.1.1 Absorption spectrum

Each substance has its own characteristic spectrum. The light absorption of the same substance at different wavelengths is different, and light absorption of different substances at the same wavelength is also different. Spectrophotometry is based on this selective absorption of

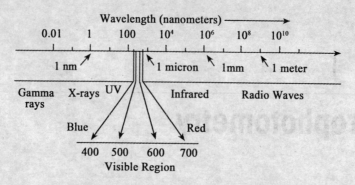

Fig. 1-1  The electromagnetic spectrum

light by a substance. Because the extent to which a sample absorbs light depends upon the wavelength of light, spectrophotometry is performed using monochromatic light.

To clearly describe the selective absorption of light by a substance, an absorption spectrum is usually plotted, which shows how the absorption of light varies with the wavelength of the light (figure 1-2). The extent of light absorption is commonly referred to as absorbance (A) or extinction (E). The absorption spectrum is a plot of absorbance vs wavelength and is characterized by the wavelength at which the absorbance is the greatest ($\lambda_{max}$). The $\lambda_{max}$ is the characteristic of each substance and provides information on the electronic structure of an analyte. Unknown substance can be identified by their characteristic absorption spectrum and $\lambda_{max}$ (qualitative analysis).

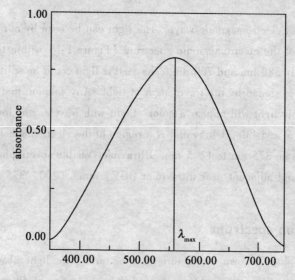

Fig. 1-2  Absorption spectrum

## 1.1.2 Basic laws of absorption of light—Lambert-Beer's Law

The quantitative analysis of spectrophotometry is based on the basic laws of absorption of light—Lambert-Beer's law. When monochromatic light (light of a specific wavelength) passes through a solution there is usually a quantitative relationship between the solute concentration and the absorption of the light, which is described by the Lambert-Beer's law. Therefore the concentration of solute can be measured by determining the extent of absorption of light at the appropriate wavelength. In order to obtain the highest sensitivity and to minimize deviations in quantitative measurement, spectrophotometric measurements are usually made using light with a wavelength of $\lambda_{max}$.

For a uniform absorbing medium the portion of the light passing through it called the transmittance, $T$. $T = I/I_0$, where $I_0$ is the intensity of the incident light, $I$ is the intensity of transmitted light. The absorbance of the light is equal to the logarithm of the reciprocal of the transmittance: $A = \lg(1/T)$.

According to Lambert-Beer's law, when a ray of monochromatic light passes through an absorbing solution, absorbance of the solution is directly proportional to the concentration of the absorbing substance and the depth of the solution through which the light passes. Equation for Lambert-Beer's law is: $A = \lg 1/T = KCL$, where $K$ is absorption coefficient (the proportionality constant that depends on the absorbing substance, wavelength of light and the temperature), $C$ is the concentration of the substance absorbing light, $L$ is the length of the path of the light. $A$ is dimensionless. When length '$L$' is in centimeter and concentration '$C$' $= 1 \text{mol/L}$, the absorbance is equal to '$\varepsilon$' (molar extinction coefficient), which is written as $\varepsilon^{1mol/L}$ and has a dimension of $1 \text{ mol/L}^{-1} \text{cm}^{-1}$. Since the molar extinction coefficient may be very large, an alternative is $E^{1\%}$, which represents the extinction given by 1cm thick sample of a 1% solution of the substance. If the $\varepsilon^{1mol/L}$ or $E^{1\%}$ is known, the amount of the substance can be quantified, $C = A/\varepsilon$, or $C = A/E^{1\%}$.

If the Lambert-Beer's law is obeyed, a plot of absorbance against concentration gives a straight line passing through the origin. Sometimes, a non-linear plot is obtained of absorbance against concentration. This is probably not satisfied with one or more of the following prerequisites of Lambert-Beer's law.

1. Light must be of a narrow wavelength range, preferably monochromatic.
2. The wavelength of light used should be $\lambda_{max}$.
3. The solution should be stable and uniform. There must not be ionization, association, dissociation of the solute during the process of spectrophotometric measurement.
4. Solution concentration is not too high ($A = 0.15 \sim 0.7$).

## 1.2 Measurement of Light Absorption

### 1.2.1 Components of spectrophotometer

A spectrophotometer is employed to measure the light absorption of a sample. There are many kinds of spectrophotometers. All the spectrophotometers employ the basic components illustrated in Figure 1-3.

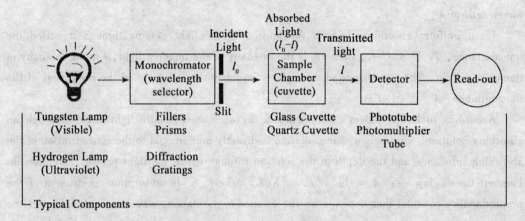

Fig. 1-3 The main components of a spectrophotometer

#### 1.2.1.1 Light sources

The source of visible light is tungsten lamp that emits light in the range of 340 to 900 nm. Some spectrophotometers employ an additional deuterium lamp emitting light in the range of 200 to 360 nm for spectral analysis in the UV range.

#### 1.2.1.2 Wavelength selector

(1) Monochromator    A monochromator is an optical device that transmits a mechanically selectable narrow band of wavelengths of light. Selection of wavelengths may be done by the use of optical filters, prisms of quartz, or diffraction grating. However, it is not feasible to have a light of a single wavelength. The monochromatic light produced by monochromator is a light that has its maximum emission at a specific wavelength, with progressively less energy at longer and shorter wavelengths. Therefore, the purer the monochromatic light is, the sensitive the measurement is.

(2) Slit system    The spectrophotometer places narrow slit in the light path that confine the light beam to a narrow path and also help to exclude light from extraneous sources. Then the intensity of the incident light can be adjusted.

#### 1.2.1.3 Sample container-cuvettes

Glass cuvette is for visible light measurement. Quartz cuvette is for UV measurement (below 340 nm). Absorbance characteristics of cuvettes should always be checked in order to obtain a standardized set for spectrophotometric measurement.

#### 1.2.1.4 Detector systems

Vacuum phototubes and photomultiplier tube are usually used to detect the transmitted light. Most detectors have a scale that reads both in absorbance units, which is a logarithmic scale, and in % transmittance, which is an arithmetic scale. The absorbance scale is normally read directly for calculation of sample concentration.

### 1.2.2 Using a spectrophotometer

Turn on the spectrophotometer and allow 15 min for warm up of the instrument prior to use. Use the wavelength knob to set the desired wavelength. With the sample cover closed, use the zero control to adjust the meter recorder to "0" on the % transmittance scale. Insert a clean cuvette containing the blank solution into the sample holder. The amount of solution placed in the cuvette is usually about 2/3 of the total volume of the cuvette. Close the cover and use the light control knob to set the meter recorder to "0" on the absorbance scale (100% transmission). Remove the blank cuvette, insert a cuvette holding the sample solution and close the cover. Read and record the absorbance. Remove the sample tube, readjust the zero, and recalibrate if necessary before checking the next sample. Carefully clean and store cuvettes for later use.

### 1.2.3 Requirements of spectrophotometric measurement

When use a spectrophotometric measurements one must understand that the absorption is produced by the particular absorbing substances (specific absorbance), but the solvent and substances in the reagents (nonspecific absorbance). The assay must include the following solutions:

(1) Blank solution/reference solution     A proper blank solution contains none of the assayed substance, but all other chemicals in the test or standard solution and undergoes the same stages as the standard and test solution. This solution will help to exclude the absorption due to reagents (nonspecific absorbance).

(2) Standard solution     It contains all the reagents of test and blank but also includes a solution of known concentration of the substance which is going to be determined in the test solution. It helps to correlate the absorption with the concentration of the concerned substance.

(3) Test solution/ sample solution/ determined solution     It contains all the reagents present in the blank and standard and undergoes the same steps, but an unknown quantity of the

concerned substance.

## 1.3 Aplications of Spectrophotometry

### 1.3.1 Qualitative analysis

Spectrophotometry can be a very useful technique for identifying unknown compounds. Absorption spectra of a pure unknown compound and a standard known compound can be generated by measuring the absorbance at a variety of wavelengths. The shape of the spectra, $\lambda_{max}$ (wavelength of maximal absorption) and $\varepsilon$ (molar extinction coefficient) can be compared to identify the property of the unknown compound. If the two spectra are completely consistent, the sample and standard compounds can be tentatively determined to have the same chromophore group, and may be the same compound. Some substances with the same chromophore group but different molecular structure may also generate the same absorption spectra, but their absorption coefficient is different. If the absorption spectra, $\lambda_{max}$ and $\varepsilon$ are complete same, they are the same material. However, for really precise qualitative analysis other assays are required.

### 1.3.2 Quantitative analysis (determining the unknown concentration)

#### 1.3.2.1 Standard addition method

The experimental approach exploits Lambert-Beer's Law, $A = KLC$, which predicts a linear relationship between the absorbance of the solution and the concentration of the analyte on the condition that $K$ and $L$ are constant. Prepare the test solution, the standard solution and the blank solution. All the solutions undergo the same treatment and their absorbances are measured under the same experimental conditions. Since the same absorbing substance, wavelength of light, temperature and length of the path of light, the $K$ and $L$ of the test and standard solutions are same. Let the concentration of the test unknown solution and standard solution is $C_u$ and $C_s$, and the absorbance of unknown solution and standard solution is $A_u$ and $A_s$, respectivly. $A_u = KLC_u$ and $A_s = KLC_s$, then $KLC_u/KLC_s = A_u/A_s$. S$_o$, $C_u$ can be calculated by the formula as following:

$$C_u = \frac{A_u}{A_s} \times C_s$$

#### 1.3.2.2 Calibration curve method

Calibration curve (also called working curve) shows how absorbance changes with the concentration of a solution. The curves are constructed by measuring the absorbance from a series of standards of known concentration and used not only to determine the concentration of an unknown sample but also to calibrate the linearity of an analytical instrument. A series of standard solutions are prepared in calibration curve method then the absorbances are measured and used to prepare a calibration curve (standard curve), which is a plot of absorbance *vs* concentration. The absorbance of the unknown solution is detected and then used to determine the concentration of the compound in the test solution in conjunction with the calibration curve.

As a perfect calibration curve, assays should normally be performed in duplicate at least while preparing standard curve and only the mean should be plotted. There must be at least five points while plotting calibration curve. The points on the calibration curve should yield a straight line. The best straight line should be drawn through the points, the origin or other points. The calibration curve may vary in different batches of reagents and hence calibration curve will be done in each. Calibration curves should never be extrapolated beyond the highest absorbance value measured. If $A$ is outside the range of the curve, the sample should be diluted or concentrated.

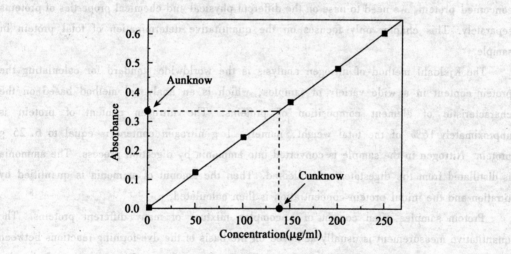

Fig. 1-4  Determining an unknown solution by calibration curve

# Experiment 1  Quantitative Analysis of Protein

Protein is the biological macromolecule which plays important roles in the cell. Protein quantitative analysis can not only help us to determine the protein content of samples but also support the diagnosis of human diseases. Proteins are composed of many amino acids linked by peptide bond. Amino acid residues, peptide bond, and some characteristics of physical and chemical properties of protein are the basis of different methods of quantitative analysis of proteins. In samples, there are many types of protein. In order to determine the quantity of the concerned protein, we need to base on the different physical and chemical properties of proteins separately. This chapter only focuses on the quantitative determination of total protein in sample.

The Kjeldahl method of nitrogen analysis is the worldwide standard for calculating the protein content in a wide variety of samples, which is an analytical method based on the characteristic of element composition of proteins. The nitrogen content of protein is approximately 16% of the total weight, namely, 1 g nitrogen content is equal to 6.25 g protein. Nitrogen in the sample is converted into ammonia by digestion process. The ammonia is distillated from the digestate and collected. Then the amount of ammonia is quantified by titration and the initial protein concentration is then calculated.

Protein samples often consist of a complex mixture of many different proteins. The quantitative measurement is usually achieved on the basis of the dye-forming reactions between functional group of the protein assayed and dye-forming reagent. The intensity of the dye correlates directly with the protein concentration and can be measured by colorimetric determination. The colorimetric protein determination methods include Folin-Phenol (Lowry) assay, BCA assay, Bradford assay and Biuret assay.

The tyrosine and tryptophan residues of proteins exhibit an ultraviolet absorbance at approximately 280 nm and the concentration of protein in a pure solution is generally proportional to the absorbance at 280 nm, so UV spectrophotometry can also be used to measure protein concentration.

In general, there is no completely satisfactory single method to determine the concentration of protein in any given sample. The choice of the method is based on the purpose, the nature and amount of protein in sample, and the nature of other components in the sample. The characteristics of the commonly used methods are summarized in the table 1-1.

Table 1-1  Methods of Protein Concentration Determination

| Method | Sensitivity range (μg/ml) | Variation between different protein | Compatibility with other detergents | Characteristics |
|---|---|---|---|---|
| The Kjeldahl method | | little | | accurate, complicated, low reproducibility |
| UV spectrophotometry | 100 ~ 1 000 | large | low | rapid, simple, recoverable |
| Biuret assay | 1 000 ~ 10 000 | little | low | high reproducibility |
| Folin-Phenol (Lowry) assay | 25 ~ 250 | large | low | convenient and inexpensive |
| Bradford assay | 10 ~ 1 000 | large | low | fast, simple, |
| BCA assay (bicinchoninic acid) | 10 ~ 1 000 | large | high | sensitive and rapid, broad linear working range |

## Protocol I    Determination of Protein Concentration by Lowry Method

### Principle

Under alkaline conditions, copper complexes with protein. When Folin-Phenol reagent (phospho-molybdic-phosphotungstic reagent) is added, it binds to the protein. Bound reagent is slowly reduced and changes color from yellow to blue.

The chromogenic principle of Folin-Phenol reagent includes the following two reactions:

① Peptide bonds in proteins + $Cu^{2+}$ in alkaline copper sulfate reagent $\xrightarrow{OH^-}$ Protein – $Cu^{2+}$ compounds

② Residues of tyrosine and tryptophan in protein-$Cu^{2+}$ compounds + Phosphomolybdic acid-phosphotungstic acid in phenol reagent $\xrightarrow{reduction}$ bluish-green compounds (molybdenum/tungsten blue)

The intensity of the blue-green color of the Folin-Phenol reaction is proportional to protein concentration. Therefore, Lowry Method, also called Folin-Phenol method, can be used to quantify proteins.

This method is simple and very sensitive. Its detectable range is 25 ~ 250 μg/ml protein. It is more subject to interference by a wide range of non-proteins, additives such as EDTA, NP-40, Triton X-100, barbital, CHAPS, cesium chloride, citrate, cysteine, diethanolamine, dithiothreitol, HEPES, mercaptoethanol, phenol, polyvinyl pyrrolidone, sodium deoxycholate, sodium salicylate and Tris. There is also much protein-to-protein variation in the intensity of color development because of different compositions of tyrosine and tryptophan in different proteins. The standard protein should be similar to the unknown. For serum, use bovine serum albumin as a standard since albumin is a major component of serum.

## Reagents

1. Alkaline copper sulfate reagent:
   Solution A: Sodium carbonate ($Na_2CO_3$) 10 g, sodium hydroxide (NaOH) 2 g and potassium/sodium tartrate (potassium saline or sodium saline) 0.25 g are dissolved in 500 ml distilled water ($dH_2O$).
   Solution B: Cupric sulfate ($CuSO_4 \cdot 5H_2O$) 0.5 g is dissolved in 100 ml $dH_2O$.
   Solution A 50 ml and Solution B 1 ml are mixed before using. The mixture is alkaline copper sulfate reagent. This reagent must be used within 24 h.
2. Phenol reagent
   Sodium wolframate ($Na_2WO_4 \cdot 2H_2O$) 100 g and sodium molybdate 25 g are added into 1.5 L flask with ground and round bottom and are dissolved in 700 ml $dH_2O$. Then, 85% phosphoric acid ($H_3PO_4$) 50 ml and concentrated hydrochloric acid (HCl) are added and mixed completely. Thereafter, the ground condensation tube is joined, reflux at small fire for 10 h. After refluxing and cooling, the condensation tube is taken off. Then, lithum sulfate ($Li_2SO_4 \cdot H_2O$) 150 g, $dH_2O$ 50 ml and several drops of bromine water are added. Continuing boiled for 15 min with the flask mouth opening in order to get rid of excessive bromine (in draft chamber). Cooling again, the solution becomes yellow or gold. (If it's still green, several drops of bromine water must be added again.) The solution is diluted to 1 L and filtrated. The filtrate is stocked in brown bottle, which is the stock solution. The acidity of the stock solution is roughly 2 mol/L. It is standardized by titrating standard NaOH (about 1 mol/L), and phenolphthalein is an indicator. It is the end point when the color of the solutions turns from red to purplish red, purplish gray, and suddenly green.
   Phenol reagent applied solution is made when the stock solution is diluted by equal volume of $dH_2O$.
3. Protein standard solution (250 μg/ml): dissolve 25 mg of crystalline bovine serum albumin in 100 ml of 0.9% NaCl.
4. 0.9% NaCl solution
5. Sample: serum.

## Procedure

1. Prepare standard curve and determine serum sample

(1) One blank solution, five standard solutions and one sample solution in duplicate are set up and operated according to the following table.

Serum (0.1 ml) is accurately taken to volumetric flask (50ml). Then 0.9% NaCl is added up to the graduation, mixed completely. Then put 0.5 ml diluted serum into the sample tube, which is processed simultaneously with standard tubes.

| Reagents (ml) | Blank | 1 | 2 | 3 | 4 | 5 | Sample |
|---|---|---|---|---|---|---|---|
| Protein standard solution | – | 0.1 | 0.2 | 0.3 | 0.4 | 0.5 | – |
| Diluted serum | – | – | – | – | – | – | 0.5 |
| 0.9% NaCl | 0.5 | 0.4 | 0.3 | 0.2 | 0.1 | – | – |
| Alkaline copper sulfate reagent | 2.5 | 2.5 | 2.5 | 2.5 | 2.5 | 2.5 | 2.5 |
| Mix and incubate 20 min at room temperature | | | | | | | |
| Phenol reagent | 0.25 | 0.25 | 0.25 | 0.25 | 0.25 | 0.25 | 0.25 |
| Equal protein concentration ($\mu g/ml$) | 0 | 50 | 100 | 150 | 200 | 250 | ? |

(2) Each tube must be mixed immediately after phenol reagent is added. (Attention: Cloudiness shouldn't appear, otherwise, chromogenic degree would be weakened.) After phenol reagent is added into the tubes for 30 min, absorbance is read at wavelength 650 nm on condition that blank tube is adjusted to zero.

2. Quantitative analysis of serum proteins

(1) Plot a standard curve of absorbance versus protein concentration of standards. According to absorbance of sample, concentration of diluted serum can be looked up in the standard curve. The concentration of original serum can be further determined by multiplying times of dilution.

(2) Serum protein concentration can also be calculated according to the following formula.

$$\text{Serum protein concentration}(g/L) = \frac{A_{sample}}{A_{standard}} \times C_{standard} \times \text{Times of dilution}$$

## Protocol II  Determination of Protein Concentration by BCA Method

### Principle

The principle of bicinchoninic acid (BCA) method is similar to Lowry method, in which $Cu^{2+}$ is first reduced to $Cu^+$ forming a complex with protein amide bonds. Then the $Cu^+$ reacts with BCA reagent to form a stable purple color compound, which is detectable at 562 nm. The absorbance value is proportional to protein concentration. Therefore, BCA method can be used to quantify proteins.

This assay has the advantage that it can be carried out as a one-step process compared to the two steps needed in the Lowry assay. Using a 96-well microplate in the experiment makes the assay quicker and easier and requires less sample volume and less BCA reagents.

## Reagents

1. BCA reagents
   Reagent A: 1 g sodium bicinchoninate (BCA), 2 g sodium carbonate, 0.16 g sodium tartrate, 0.4 g NaOH, and 0.95 g sodium bicarbonate, brought to 100 ml with distilled water. Adjust the pH to 11.25 with 10 M NaOH.
   Reagent B: 0.4 g Cupric Sulfate·$5H_2O$, in 10 ml distilled water.
   Standard working solution: Mix reagent A and reagent B according to volume ratio of 50:1. The working solution is stable for 1 week and should be green.
2. BSA protein standard solution (1 mg/ml): dissolve 10 mg of crystalline bovine serum albumin (BSA) in 10 ml of 0.9% NaCl.
3. 0.9% NaCl solution.
4. Sample: serum. Prepare serial dilutions of serum with 0.9% NaCl solution.

## Procedure

1. Make a dilution series of BSA protein standard solution to prepare a set of protein standards of known concentration (0.1 mg/ml, 0.2 mg/ml, 0.4 mg/ml, 0.8 mg/ml, 1.0 mg/ml).
2. Pipette 20 μl of each standard or sample in duplicate (or triplicate) wells in a microtiter plate, including blanks lacking protein.
3. Add 200 μl working reagent to each well.
4. Mix well by pipetting up and down several times. Do not introduce any bubbles into the solution.
5. Incubate at 37℃ for 30 min. Cool samples before reading the absorbance.
6. Determine the absorbance at 562 nm using a spectrophotometric microplate reader. Subtract the average blank from the measurements of all other standards and samples.
7. Calculate concentration of serum.
(1) Plot a standard curve of absorbance versus protein concentration of standards. According to absorbance of sample, concentration of diluted serum can be looked up in the standard curve. The concentration of original serum can be further determined by multiplying times of dilution.
(2) Serum protein concentration can also be calculated according to the following formula.

Chapter I  Spectrophotometry

$$\text{Serum protein concentration}(g/L) = \frac{A_{sample}}{A_{standard}} \times C_{standard} \times \text{Times of dilution}$$

## Protocol III  Determination of Protein Concentration by Bradford Method

### Principle

Bradford method also called coomassie brilliant dye-binding method. In acidic solution, coomassie brilliant blue (CBB) G250 could combine with protein, and then it shows a shift in its absorption maximum from 465 nm to 595 nm, and its colour changes from red to blue. The absorption at 595 nm is directly proportional to protein concentration in a definite range of 1 to 1000 μg/ml. So we can determine protein concentration by this method.

This method is easy to perform, quick, highly sensitive, stable and has less interfering substances. It is a widely used method, especially for determining protein content of cell fractions and samples for gel electrophoresis.

### Reagents

1. 0.9% NaCl solution.
2. 100 μg/ml protein standard solution: dissolved 10 mg bovine serum albumin in 100 ml 0.9% NaCl solution.
3. Coomassie brilliant blue G250 solution: dissolve 0.1g Coomassie brilliant blue G250 in 50 ml ethanol (95%), then add 100 ml 85% (m/v) phosphoric acid. Dilute to 1 L when the dye has completely dissolved and filter through Whatman #1 paper just before use. This solution can be preserved for 1 month at room temperature.
4. Sample solution: serum. Prepare serial dilutions of serum with 0.9% NaCl solution.

### Procedures

1. Prepare standard curve and determine serum sample

One blank solution, five standard solutions and one sample solution in duplicate are set up and operated according to the following table.

Mix the solution and allow the tubes to stand at room temperature for 10 min. Determine absorbance at 595 nm on condition that blank tube is adjusted to zero. The used cuvettes should be immediately washed by 95% alcohol after the measurement.

| Reagents (ml) | Blank | 1 | 2 | 3 | 4 | 5 | sample |
|---|---|---|---|---|---|---|---|
| Protein standard solution | – | 0.2 | 0.4 | 0.6 | 0.8 | 1.0 | – |
| Diluted serum | – | – | – | – | – | – | 1.0 |
| 0.9% NaCl solution | 1.0 | 0.8 | 0.6 | 0.4 | 0.2 | – | – |
| Coomassie bright blue G250 | 5.0 | 5.0 | 5.0 | 5.0 | 5.0 | 5.0 | 5.0 |
| Equal protein concentration ($\mu g/ml$) | 0 | 20 | 40 | 60 | 80 | 100 | ? |

2. Calculate concentration of sample

(1) Plot a standard curve of absorbance versus protein concentration of standards. According to absorbance of sample, concentration of sample can be looked up in the standard curve.

(2) Serum protein concentration can also be calculated according to the following formula.

$$\text{Serum protein concentration}(g/L) = \frac{A_{sample}}{A_{standard}} \times C_{standard} \times \text{Times of dilution}$$

## Protocol IV    Determination of Protein Concentration by Biuret Method

### Principle

Under alkaline conditions, the compounds containing two or more peptide bond (-CONH-) react with copper sulfate in $Cu^{2+}$ forming a purple complex which can be measured at 540 nm. This reaction is called biuret reaction. Protein contains many peptide bonds and so it can be determined based on the biuret reaction.

Biuret method is fast, convenience and has very few interfering agents such as ammonium salts. It is the most widely used method for total protein determination. However, this assay has a relatively low sensitivity and a low specificity (biuret reagent could also react with the compounds containing groups such as $-CONH_2$、$-CH_2NH_2$、$-CS-NH_2$).

### Reagents

1. 6 mol/L NaOH: Dissolve 60 g NaOH in 250 ml distilled water.
2. Biuret reagent: Dissolve $CuSO_4 \cdot 5H_2O$ 2.5 g, potassium tartrate 10.0 g and potassium iodide 5.0 g, and added in 100 ml 6 mol/L NaOH, constant volume to 1 L. Store in brown container. Discard if a black precipitate forms.
3. 0.9% NaCl solution.
4. 10 mg/ml protein standard solution: dissolved 500 mg bovine serum albumin (BSA) in 50 ml 0.9% NaCl solution.
5. Sample solution: serum. Prepare serial dilutions of serum with 0.9% NaCl solution.

## Procedure

**1. Prepare standard curve and determine serum sample**

One blank solution, five standard solutions and one sample solution in duplicate are set up and operated according to the following table.

| Reagents (ml) | Blank | 1 | 2 | 3 | 4 | 5 | sample |
|---|---|---|---|---|---|---|---|
| Protein standard solution | - | 0.2 | 0.4 | 0.6 | 0.8 | 1.0 | - |
| Diluted serum | - | - | - | - | - | - | 1.0 |
| 0.9% NaCl solution | 1.0 | 0.8 | 0.6 | 0.4 | 0.2 | - | - |
| Biuret reagent | 4.0 | 4.0 | 4.0 | 4.0 | 4.0 | 4.0 | 4.0 |
| Equal protein concentration (mg/ml) | 0 | 2 | 4 | 6 | 8 | 10 | ? |

Mix the solution and allow the tubes to stand at room temperature for 30 min. Determine absorbance at 540 nm on condition that blank tube is adjusted to zero.

**2. Calculate concentration of sample**

(1) Plot a standard curve of absorbance versus protein concentration of standards. According to absorbance of sample, concentration of sample can be looked up in the standard curve.

(2) Serum protein concentration can also be calculated according to the following formula.

$$\text{Serum protein concentration}(g/L) = \frac{A_{sample}}{A_{standard}} \times C_{standard} \times \text{Times of dilution}$$

## Protocol V  Determination of Protein Concentration by UV spectrophotometry

### Principle

The ultraviolet absorption characteristic of proteins depends on the Tyr and Trp content (and to a very small extent on the amount of Phe and disulfide bonds). The absorption value at 280 nm is directly proportional to the protein content. Different proteins have widely varying content of Tyr and Trp, therefore the $A_{280}$ varies greatly between different proteins. Unlike the colorimetric process, this method is less sensitive and requires higher protein concentrations and is suitable for concentration determination of pure protein solutions when a standard protein with similar content is available.

The advantages of this method are that it is simple, and the sample is recoverable. But

$A_{280}$ may be disturbed by the parallel absorption of non-proteins (e.g. DNA). The maximal ultraviolet absorption of nucleic acids is at 260 nm. Therefore by measuring the absorption at 280 nm and 260 nm, it is possible to correct for the nucleic acid present and roughly estimate protein concentration in a sample using an empirical formula. However, the absorbance at 260 nm and 280 nm are variable in different proteins and nucleic acids that errors may be caused in measurement.

The specific extinction coefficient of a number of proteins at 280 nm has been determined, so the amount of the pure samples can also be quantified using the known $\varepsilon^{1mol/L}$ or $E^{1\%}$.

## Reagents

1. Distilled water ($dH_2O$).
2. 10 mg/ml protein standard solution: dissolved 500 mg bovine serum albumin (BSA) in 50 ml 0.9% NaCl solution.
3. Sample: protein solution with unknown concentration.

## Procedures

1. Mix 1 ml of standard or sample solution with 2 ml of $dH_2O$ and measure the absorbances (A) at wavelength 280 nm and 260 nm respectively using quartz cuvette with 1 cm light path, on condition that absorbance of $dH_2O$ is adjusted to zero.
2. Quantitative analysis of sample

   (1) Read $A_{280}$ and $A_{260}$ of the sample, calculate protein concentration using the following formula.

   Lowry-Kalckar formula: Protein concentration (mg/ml) = $1.45A_{280} - 0.74A_{260}$

   Warburg-Christian formula: Protein concentration (mg/ml) = $1.55A_{280} - 0.76A_{260}$

   (2) If the sample is a pure protein with known $\varepsilon^{1mol/L}$ or $E^{1\%}$ at 280 nm, the amount of the substance can be quantified using the following formula:

   $$C = A/\varepsilon, \text{ or } C = A/E^{1\%}$$

   Concentration unit is g/dL, or mol/L, depending on which type of extinction coefficient is used.

Chapter I  Spectrophotometry

**Experiment title**
**Date**
**Observations and results**

**Discussion**
1. Absorption at 280 nm by proteins depends on the _____ and _____ content.
2. What are the colorimetric methods used to determine the concentration of serum protein? Describe briefly the advantages and disadvantages of each method.

**Teacher's remarks**
**Signature**
**Date**

Experimental Manual in Medical Biochemistry

## Experiment 2  Determination of Blood Glucose

### Part I  Modified Ortho-Toluidine in Boric Acid method (O-TB method)

### Principle

Glucose is hexose containing an aldehyde group. Heated up under the acidic condition, it dehydrates to be 5-hydroxyl-methyl—2-furfural, which condenses with ortho-toluidine to produce a stable green compound. Its color intensity is directly proportional to glucose content of the sample. Glucose contents of samples can be determined by compared with standard glucose solution assayed by the same method.

### Reagents

1. Saturated boric acid solution: 6 g boric acid is dissolved in 100 ml $dH_2O$, place it overnight and then filter it.
2. O-Toluidine reagents: 1.5 g thiourea is dissolved in 883.2 ml acetic acid and mixed with 76.8 ml ortho-toluidine. Then add 40 ml saturated boric acid solution into it. The reagents may be stored in brown reagent bottle for several months.
3. Stock glucose standard solution: 10.0 g anhydrous glucose is dissolved in 1 L 0.25 % benzoin acid.
4. Working glucose standard solution: 1.0 mg/ml working glucose standard solution is prepared by ten times dilution of glucose standard stock solution with 0.25% benzoin acid.

### Procedure

1. Three test tubes are used and the operation is done according to the following table.

| Reagent (ml) | Blank tube | Standard tube | Test tube |
| --- | --- | --- | --- |
| $dH_2O$ | 0.1 | 0.0 | 0.0 |
| Blood plasma | 0.0 | 0.0 | 0.1 |
| Glucose standard solution | 0.0 | 0.1 | 0.0 |
| O-TB reagent | 5.0 | 5.0 | 5.0 |

2. After mixing the tubes, heat the tubes at 100° C (boiling water) for 5~6 min and then cool them in running water. The absorbance is read within 30 min at wavelength 620 nm under the condition that absorbance is adjusted to zero with the blank tube. Samples with a very high concentration will produce a very high absorbance which can not be read accurately on a spectrophotometer. Their absorbances can be measured again after being diluted with O-Toluidine reagents.
3. Blood glucose concentration can also be calculated according to the following formula.
Blood glucose levels(mg/dL) = $A_{Test}/A_{standard} \times C_{standard} \times 100$

## Clinical significance

1. Glucose is a major sugar in blood. Blood glucose levels fluctuate physiologically as well as in diseased condition. Therefore, glucose measurement is a very important assay. It is most commonly performed in the detection and treatment evaluation of diabetes.
2. Normal blood glucose level in fasting serum/ plasma is 70~110 ml/dL or 3.89~6.11 mmol/L. A condition in which an excessive amount of glucose circulates in the blood is referred as hyperglycemia. And hypoglycemia refers to a pathologic state produced by a lower than normal blood glucose level.
3. The concentrations of glucose will increase in some patients, who suffer from endocrine disease such as DM (diabetes mellitus), endocrine diseases (hyperthyroidism, hypercorticism). Moderate rise of glucose level is associated with infectious diseases, intercranial diseases like meningitis, encephalitis, tumors and hemorrhage. Anesthesia also causes increase in glucose level. Physiological hyperglycemia occurs due to the take-up of high sugar food, or the increased secretion by adrenal gland because of tight mood.
4. Hypoglycemia is due to overdose of insulin or oral hypoglycemic agents or a variety of hormonal causes. Low levels of glucose under fasting conditions are associated with hypothyrodism, hypoadrenalism, hypopituitarism and also in glycogen storage disease. The concentration will be low in some long-term malnutrition people or the men who can not eat. Physiological hypoglycemia may occur due to starvation or violent exercise.

## Part II  Effects of insulin and adrenalin on blood glucose level

### Principle

Blood glucose levels in human are mainly regulated by hormones. Insulin reduces blood glucose but the other hormones such as adrenalin can increase blood glucose. In this experiment, insulin is injected into one rabbit and adrenalin is injected into the other. The venous blood is collect from the two rabbits before and after injecting and the blood glucose levels of the samples are determined. Analyze the variations of blood sugar before and after injection and learn the effects of insulin and adrenalin on the concentration of blood glucose.

## Reagents

1. Potassium oxalate.
2. 25% glucose.
3. Adrenalin.
4. Insulin.
5. Dimethylbenzene.

## Procedure

1. Prepare experimental animals: select two normal rabbits, take their weights after fasting overnight.
2. Collect blood samples before injection: collect the blood from ear-border vein. Hairs around ear-border vein are firstly removed. Clean the skin and make the vein hyperaemic with dimethylbenzene. Then the vein is poked with a blade. Venous blood is collected into blood-collection tubes containing 2 mg/ml final concentration of potassium oxalate and the tubes are shaked while collecting to prevent the blood from coagulating. Stop bleeding with cotton by pressing blood vessel. Finally, isolate plasma from the collected whole blood.
3. Inject hormones: one rabbit is subcutaneously injected insulin at a dosage of 0.75 units/kg body weight. The other one is injected adrenalin at a dosage of 0.4 mg/kg body weight. The time of injection are recorded respectively.
4. Collect blood samples after injection: follow the same operations as in step 2. The time for blood collection: 30 min after adrenalin injection and one hour after insulin injection. After blood collection, inject subcutaneously 10 ml 25% glucose to prevent the rabbit from insulin shock (hypoglycemia shock).
5. Measure blood glucose level of the samples: the method is the same as in "Part I".
6. Analyze the effects of hormones: calculate the increased and decreased percentage of blood glucose after the injection of insulin and adrenaline respectively.

## Clinical significance

1. Insulin decreases glucose level

    Insulin is the only hormone to reduce the blood glucose level. It enhances the uptake of glucose into the liver, stimulates glycogen synthesis and glycolysis, and inhibits gluconeogenesis.

2. Adrenalin increases glucose level

    Adrenaline is a stress hormone, which is secreted as a result of stressful stimuli (fear, excitement, hemorrhage, hypoxia, hypoglycemia, etc) and leads to glycogenolysis in liver and muscle. It blocks insulin secretions, impairs insulin action in target tissues, so that hepatic glucose production is increased and the capacity to dispose exogenous glucose is impaired.

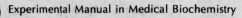

**Experiment title**
**Date**
**Observations and results**

**Discussion**

1. Which hormones elevate or reduce blood glucose?
2. Why does the blood glucose level increase as a result of stressful stimuli?

**Teacher's remarks**
**Signature**
**Date**

# Chapter II
# Electrophoresis

Electrophoresis, one kind of electrokinetic phenomenon first discovered by Reŭss in 1809, has been developed as a typical biochemical technique and universally practiced throughout the fields of biology, biochemistry and molecular biology. Some theoretic knowledge including basic principles, impact factors, detection of separated components and applications of special electrophoresis, and common experiments of electrophoresis applying in medical laboratories are introduced in this chapter.

## 2.1 Basic Principles of Electrophoresis

Electrophoresis is the forced migration of charged particles in an electric field. Cations move toward the cathode and anions move toward the anode. An electrophoretic system consists of a voltage power supply, electrodes, buffer, and a supporting medium such as filter paper, cellulose acetate strips, polyacrylamide gel, or a capillary tube. Electrophoresis is a simple, rapid, and sensitive technique to separate and purify the charged biomolecules including amino acids, proteins, and nucleic acids.

When an electric field is applied, a charged particle will experience a forward electric force "$F$", and a reverse frictional force "$f$" due to frictional resistance of the matrix. The electric force is proportional to the net charge on the molecule, $Q$, and the electric field strength, $E$.

$$F = EQ$$

If electrophoresis occurred in free solution rather than within a gel, the frictional force of a globular molecule would follow Stokes' law:

$$f = 6\pi r v \eta$$

Where $r$ is the radius of a particle, $\eta$ is the viscosity of the medium, and $v$ is the migration velocity of the particle. When the two forces, $F$ and $f$ reach equilibrium ($EQ = 6\pi r v \eta$), the particle will migrate in the electric field with a constant velocity:

$$v = EQ/6\pi r \eta$$

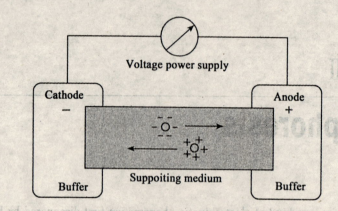

Fig. 2-1 An electrophoretic system

The migration velocity is usually represented by the electrophoretic mobility, $M$.

$$M = v/E = Q/6\pi r\eta$$

In an electrophoretic system, molecules have different migration mobilities depending on their total charge, size and shape, and therefore can be separated after a given time.

Stoke's law, however, is not sufficient to describe the frictional force of a particle within a gel matrix. In addition to the viscosity of medium, the frictional resistance is also determined on the density and effective "pore size" of the gel. The result of this combination of factors is that among molecules with the same charge to mass ratio, larger molecules move slowly in gel electrophoresis and electrophoretic separation occurs by size. In techniques such as denaturing DNA/RNA agarose gel electrophoresis or protein SDS-polyacrylamide gel electrophoresis (SDS-PAGE), conditions are maintained so that the charge to mass ratio is virtually equal among sample molecules and separation occurs based on the molecular size.

## 2.2 Impact Factors of Electrophoresis

### 2.2.1 Properties of the molecules

#### 2.2.1.1 Molecular charge

Sample molecules, a mixture of proteins for example, have different net charges depending on the pI of molecule and the pH of the electrophoresis buffer. When an electric field is applied, charged molecules migrate in the direction of the electrode bearing the opposite charge. Namely, the positively charged molecules migrate towards the cathode and the negatively charged molecules migrate toward the anode of an electrophoretic apparatus. The more charges they have, the faster they migrate. Hence, proteins can be separated based on their charges.

#### 2.2.1.2 Molecular shape

Molecules can be separated by gel electrophoresis because of their difference in shape. Compared with globular proteins, the long and loose proteins tend to interact with the gel network more tightly and therefore travel more slowly.

#### 2.2.1.3 Molecular size

The molecules with different size can be separated by electrophoresis, regardless their same charge and shape. Of course, a smaller one has a faster electrophoretic mobility.

### 2.2.2 Properties of the electrophoretic system

#### 2.2.2.1 Electric field strength

The mobility of molecules can be improved by increasing the strength of electric field. However, the electric current and heat are increased subsequently, which may cause damages of the molecules. So, an efficient cooling system is essential in a high-voltage electrophoresis.

#### 2.2.2.2 Support medium

Common zonal support media used in electrophoresis are paper and starch, cellulose acetate membrane (CAM), agarose gel, and polyacrylamide gel. Paper electrophoresis is one of the earliest forms of zone electrophoresis, in which a strip of filter paper is used as a medium to support a thin layer of buffer. CAM give sharper bands and better resolution than paper, bind proteins less and have less electroendosmosis. At present, CAM electrophoresis replaces paper electrophoresis and remains in use, mainly in medical diagnostic laboratories. Agarose is a polysaccharide obtained from seaweed. Agarose gel has a macroreticular structure, and zero electroendosmosis. It can be cast onto a sheet of flexible plastic Gel-Bond and it can be dried to provide a durable record. Polyacrylamide gel has the advantage of being a synthetic gel, which is highly reproducible. Moreover, the pore size can be controlled by varying the proportions of acrylamide and the crosslinking agent, bisacrylamide. The microreticular nature of polyacrylamide gels has sieving effect which greatly increases its resolution and sensitivity. Support media may influence separation in three aspects: restrictions on mobility, electroendosmosis and effect of molecular sieve.

(1) **Electroendosmosis**: The support medium and/or the surface of the separation equipment, such as glass plates, tubes or capillaries can carry charged groups (e.g. carboxylic groups in starch and agarose, sulfonic groups in agarose, silicon oxide on glass surfaces). In basic and neutral buffers, these groups ionize to negatively charged ions and will be attracted by the anode when an electric field is applied. Because these negatively charged groups are immobilized, so that a counter-flow of $H_3O^+$ (hydronium) ions towards the cathode will happen. This effect is electroendosmosis and may result in blurred zones and distort the migration of the samples.

**(2) Effect of molecular sieve**: Gel is a porosity support medium, which can act as a sieve to improve resolution by reducing diffusion broadening of bands. The larger the molecule is, the slower it migrates. Pore sizes in the gels can be controlled by regulating the concentration of the monomer units of the gel, depending on the samples being separated.

### 2.2.2.3 Properties of buffer

The pH value of the buffer influences the charge density of a molecule, and will influence the migrating direction and velocity of the molecule consequently. In vertical or capillary systems, the pH is very often set to a very high (or low) value, so that as many as possible sample molecules are negatively (or positively) charged, and thus migrate in the same direction.

Ionic Strength also influences rate of separation. The weaker is the ionic strength, the faster is the migration velocity of charged particles. However, if the ionic strength is too low, the buffering ability of the buffer will also decrease. This will increase the diffusion of samples and decrease the resolution of electrophoresis. Ionic strength (I), can be calculated from the following equation:

$$I = \frac{1}{2} \sum c_i z_i^2 ,$$

where the sum is over all ionic species of concentration $C_i$ and valence $Z_i$. Normally, the most suitable ionic strength is about 0.02 to 0.2.

## 2.3 Detection of Components after Electrophoretic Separation

Many macromolecules can not be observed directly, different samples including proteins and nucleic acids should be stained by corresponding methods.

### 2.3.1 Detection of proteins

#### 2.3.1.1 Staining by special protein dye

Chemical dyes such as amido black-10B and Coomassie brilliant blue (CBB) R250 or G250 are commonly used for staining and visualization of proteins. A more sensitive staining method, silver staining is commonly used for detecting proteins in polyacrylamide gel electrophoresis.

#### 2.3.1.2 Detection by fluorescence labeling

Fluorescent dyes (Cy2, Cy3 and Cy5) can be used to bind the amino groups of proteins, which are detected by analysis of the relative fluorescence intensities.

#### 2.3.1.3 Detected by radiolabeling

Isotopes $^{14}C$, $^{3}H$, and $^{32}P$ are commonly used to label proteins, which can be then detected with an auto-radiographic analysis.

#### 2.3.1.4 Immunochemical detection

In Western blotting, the protein on the membrane can be recognized and bound with its specific antibody, and the corresponding second antibody, which is conjugated with an enzyme such as horseradish peroxidase (HRP). This is visualized by a colorimetric reaction catalyzed by HRP. Immunochemical detection is usually used to identify a specific protein in a mixture.

### 2.3.2 Enzymes

Appropriate enzymatic methods will be used, in which the products or substrates of the catalyzed reaction can be detected by an instrument.

### 2.3.3 Detection of glycoproteins

Schiffs Reagent and Alcian blue is used for staining and detecting neutral glycoprotein and acidic mucopolysaccharides, respectively.

### 2.3.4 Lipid detection

Sudan Black is a dye commonly used to stain lipids.

### 2.3.5 Staining of nucleic acids

Ethidium Bromide (EB) and SYBR green or gold are commonly used to bind DNA and visualized under UV-light.

## 2.4 Special Electrophoretic Techniques and Applications

Based on the development of support media and basic principle of electrophoresis, many kinds of electrophoretic techniques have been generated and wildly used in life science researches. Some special electrophoretic techniques are introduced as following.

### 2.4.1 Isoelectric focusing (IEF)

Isoelectric focusing (IEF) is an electrophoretic technique that has been widely used for the separation of proteins based on differences in their isoelectric points (pI). IEF takes place in a pH gradient, which is formed by special amphoteric buffers (ampholytes) on an electric field. When an electric field is applied, the negatively and positively charged ampholytes move towards the anode and cathode respectively, and a stable gradually increasing pH gradient depending on the initial mixture of ampholytes is formed. The ampholytes are generally provided in a support matrix, a polyacrylamide gel or an agarose gel. During the electrophoresis, proteins or peptides migrate through a stable pH gradient until they reach the pH equal to their pI, where the net charge and mobility of proteins are zero. In case of diffusion to an adjacent pH-environment, molecular particles will rapidly acquire a charge and move back again and will 'focus' at their pI site.

Because pore sizes of polyacrylamide gels can be accurately controlled by regulating the total acrylamide concentration and degree of cross-linking (relationship between acrylamide and bis-acrylamide), the polyacrylamide gel is more commonly used in IEF to separate some charged macromolecules, such as proteins and isoenzymes. Commercial immobilized pH gradient (IPG) strips with different pH ranges are available, which facilitate the application of IEF. IEF is also used in some complex separation techniques to increase separation efficiency, such as two-dimensional polyacrylamide gel electrophoresis (2D-PAGE, or 2DE), in which IEF is coupled with SDS-PAGE, and capillary isoelectric focusing (CIEF), in which IEF is combined with a capillary electrophoresis.

### 2.4.2 Capillary electrophoresis (CE)

Capillary electrophoresis (CE) is a special electrophoretic technique that performs electrophoresis by using narrow-bore fused-silica capillaries to separate a complex array of large and small molecules. The capillary tube has a high surface to volume ratio, so that it easily radiates heats and prevents the denaturation of biomolecules including proteins and enzymes during the high-voltage electrophoresis. Applying injection of samples to the capillary tubes and output of the detection digital to a computer is of great benefit to the automation of CE.

The capillary tubes can be filled with different matrices according to different samples. Depending on the type of capillaries and electrolytes used, CE can be classified into several types. For example: capillary zone electrophoresis (CZE), which is based on differences in the charge-to-mass ratio of analytes; capillary gel electrophoresis (CGE), which is the adaptation of traditional gel electrophoresis into the capillary using polymers in solution to create a molecular sieve, and commonly used for DNA sequencing and genotyping; capillary isoelectric focusing (CIEF) that allows amphoteric molecules to be separated by electrophoresis in a pH gradient generated between the cathode and anode, and is commonly used to determine a protein's pI.

### 2.4.3 Two-dimensional polyacrylamide gel electrophoresis(2D-PAGE, or 2DE)

Two-dimensional polyacrylamide Gel Electrophoresis is a special complex electrophoresis of IEF and SDS-PAGE, named as 2D-PAGE or 2DE, and is the most popular protein separation technique in proteomics study. The separation of proteins is performed with the first dimension IEF according to their pI, and the second dimension SDS-PAGE according to their molecular weights (MW). Usually, proteins are subjected to SDS-PAGE in a perpendicular direction of IEF. After separation in first dimension of IEF using a long thin immobilized pH gradient (IPG) strip, proteins are stained as bands. The IPG strip is then inserted into the gel cassette on top of an SDS-PAGE slab gel. When the second dimension electrophoresis is finished, the protein pattern, in the gel is a number of spots rather than bands (Figure 2-2).

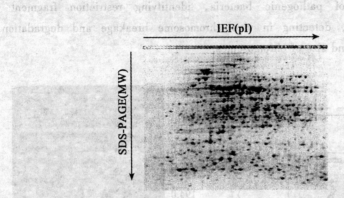

Fig. 2-2 A typical results of 2DE

At present, 2D-PAGE is a powerful tool of proteomics research to analyze differences of protein expressional profiles. Traditionally, 2D-PAGE gels are stained using Coomassie brilliant blue (CBB) or the more sensitive silver staining procedures, and analyzed using computer software for changes in protein expression. 2-dimensional difference gel electrophoresis (2D-DIGE) is a modified 2D-PAGE method, in that each sample is pre-labeled with a fluorescent dye prior to IEF. Three dyes (Cy2, Cy3 and Cy5) are commercially available and therefore up to 3 samples can be run on the same gel at any one time. Analysis of 3 samples on one 2D-DIGE gel rather than three individual 2D-PAGE gels will reduce the experimental gel-to-gel variation.

### 2.4.4 Pulsed-Field Gel Electrophoresis (PFGE)

Pulsed-field gel electrophoresis (PFGE) is a technique used to separate especially long strands of DNA (10~2000 kb). During common agarose gel electrophoresis, the electric field is continuous and DNA fragments above 30 kb migrate with the same mobility regardless of size, lead to a single large diffuse band. In PFGE, the DNA is forced to change direction during electrophoresis by pulsed alternation of electric fields; different sized fragments (within the diffuse band in common agarose gel electrophoresis) begin to separate from each other. Specialized equipment is required to perform PFGE, which is at least consisting of a gel rig with clamped electrodes in a hexagonal design, a chiller, a pump, and a programmable power supply.

PFGE separation effect is associated with some parameters, such as the applied voltage and field strength, pulse length, the agarose concentration, the buffer chamber temperature, and the amount of DNA loaded. Usually, the longer the pulse time, the larger the DNA fragments to be suitable for separation.

PFGE has been widely used for gene mapping and medical epidemiology studies, such as epidemiological typing of pathogenic bacteria, identifying restriction fragment length polymorphisms (RFLP), detecting in vivo chromosome breakage and degradation, and determining the number and size of chromosomes.

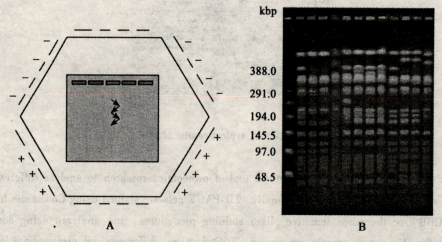

Fig. 2-3  A: Schematic gel system of the PFGE.  B: Typical results of PFGE

## Experiment 3  Separation of Serum Proteins by Cellulose Acetate Membrane Electrophoresis

### Principle

Separation and detection of serum/plasma proteins is a routine work in medical diagnostics. The most common method is the electrophoresis by using a strip of cellulose acetate membrane (CAM) as the supporting medium. Since the membrane served only to support the buffer, CAM electrophoresis can be considered as a form of free electrophoresis. Serum proteins are negatively charged in alkaline buffer, so they will migrate toward anode in an electric field. Because of the differences of net charge and molecular weight, their migration velocities are distinct. Proteins separated on film are fixed in position and stained with a protein-specific dye, such as amido black. After destaining, five zones indicating fractions of albumin, $\alpha_1$-globulin, $\alpha_2$-globulin, $\beta$-globulin, and $\gamma$-globulin of serum proteins can be visualized. Their percentages can be estimated by scanning with densitometry. A typical result is shown in figure 2-4.

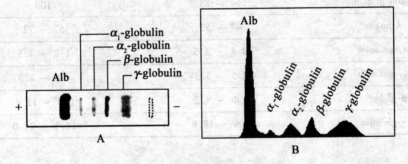

Fig. 2-4  A: Pattern of separated serum proteins visualized in the CAM strip.
B: Densitometric scanning of the CAM strip converts bands to characteristic peak of 5 fractions.

### Reagents

1. The electrophoresis buffer: 0.05 mol/L barbital sodium / barbital (pH 8.6). Take 12.76 g barbital sodium and 1.66 g barbital and dissolve them in 1 L distilled water ($dH_2O$).
2. Staining solution: take 0.5 g amido black 10B, dissolve it in 40 ml $dH_2O$, 50 ml methanol and 10 ml glacial acetic acid.

3. Destaining solution: 45 ml 95% alcohol, 5 ml glacial acetic acid, and 50 ml $dH_2O$.

## Procedure

1. Immerse a 10 cm × 2 cm cellulose acetate membrane (CAM) strip into the electrophoresis buffer, take it out and absorb water on the membrane with filter paper.
2. Add 3~5 μl sample on the rough side of membrane at the site 2 cm from cathode, and put the membrane on the electrophoresis device with the rough side of membrane down, connect the electrode and the membrane with filter paper. Turn on the power and electrophoresis is carried out at a constant voltage of 10 V/cm. The electrophoresis continues for 50~60 min.
3. After electrophoresis, take out CAM strip, immerse into the dye solution for 5 min.
4. Put the strip into the destaining solution, shaking the strip with a forceps for rinsing, and repeat three times till the background of the membrane are colorless.
5. Observe and judge the shade of color and width of the bands by naked eyes to estimate their quantity roughly or estimate their percentages by scanning with densitometry.

## Reference

Table 2-1    Reference values of serum proteins

| Proteins | Concentration (g/L) | Percentage (%) |
|---|---|---|
| Total protein | 64 ~ 83 | 100 |
| Albumin | 43 ~ 53 | 57 ~ 72 |
| $\alpha_1$-globulin | 0.4 ~ 2.6 | 1.8 ~ 4.5 |
| $\alpha_2$-globulin | 2.5 ~ 5.3 | 4.0 ~ 8.3 |
| β-globulin | 4.0 ~ 8.2 | 6.8 ~ 11.4 |
| γ-globulin | 7.6 ~ 18.6 | 11.2 ~ 23.0 |

## Clinical significance

The serum proteins are actually a very complicated mixture that includes not only simple proteins but also mixed or conjugated proteins such as glycoproteins and various types of lipoproteins. The fractions of the serum proteins separated by CAM electrophoresis, especially $\alpha_1$-globulin, $\alpha_2$-globulin, β-globulin, and γ-globulin are mixtures. Albumin is the major protein of human serum produced by liver about 12 g per day, it has two main functions: ① maintaining colloid osmotic pressure of blood; ② transportation: acting as a carrier molecule for bilirubin, fatty acids, trace elements and many drugs. The significant changes of the total and fractions of serum proteins are associated with some diseases.

(1) Decreased total protein may indicate malnutrition, nephrotic syndrome, and gastrointestinal protein-losing enteropathy.

(2) Increased $\alpha_1$-globulin may be seen in rheumatoid arthritis, systemic lupus erythematosus (SLE), malignancy and acute inflammatory disease; decreased $\alpha_1$-globulin may see in $\alpha$-1 antitrypsin deficiency.

(3) Increased $\alpha_2$-globulin may appear in acute inflammation, nephrotic syndrome, and chronic inflammation; decreased $\alpha_2$-globulin may appear in hemolysis.

(4) Increased $\beta$-globulin may happen in myeloma, estrogen therapy, and hyperlipoproteinemia (such as familial hypercholesterolemia); decreased $\beta$-globulin may happen in congenital coagulation disorder, consumptive coagulopathy, disseminated intravascular coagulation.

(5) Increased $\gamma$-globulin may be the results of multiple myeloma, rheumatoid arthritis, SLE, chronic hepatitis, hyperimmunization, and Waldenstrom's macroglobulinemia.

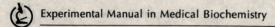

**Experiment title**
**Date**
**Observations and results**

**Discussion**

1. The support medium of electrophoresis influences separation in _____, _____, _____, _____ three aspects.
2. Serum proteins were separated by cellulose acetate membrane electrophoresis into five zones: _____, _____, _____, _____, _____, respectively.
3. What are the functions of serum proteins?

**Teacher's remarks**
**Signature**
**Date**

## Experiment 4  Separation of Serum Lipoproteins by Agarose Gel Electrophoresis

### Principle

In alkaline buffer solution, serum lipoproteins are negatively charged. They will migrate toward anode in electric field. Because of the difference in their compositions, the lipoprotein particles have different charges and sizes, which lead to their different migration velocities. Serum lipoproteins can be separated into at least four kinds of bands by agarose gel electrophoresis, including α-lipoproteins, pre-β-lipoproteins, β-lipoproteins and chylomicrons (CM). The four major bands are related to the different lipoprotein layers obtained by ultracentrifugation (Figure 2-5).

Prestaining with a lipid dye Sudan black B, the bands of serum lipoproteins can be seen during electrophoresis. The results of percentages of lipoproteins are estimated as the shade of color and width of the bands by naked eye, or calculated by scanning with a spectrophotometer.

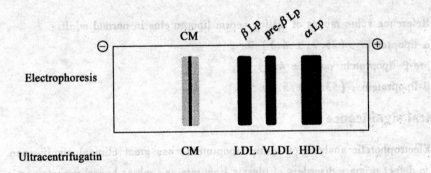

Fig. 2-5  Pattern of separated serum lipoproteins

### Reagents

1. The electrophoresis buffer: Take 15.5 g barbital sodium and dissolve it in distilled water, add 1 mol/L HCl to adjust pH to 8.6, and make the final volume up to 1 L with distilled water.
2. The agarose gel buffer: Take 10.3 g barbital sodium and dissolve it in distilled water, add 1 mol/L HCl to adjust pH to 8.6 and make the final volume to 1 L with distilled water.

3. Dye solution: Dissolve 1g Sudan Black B in 100 ml Petroleum ether/ethanol (1:4, *V/V*) solution.
4. 0.8% agarose gel: Take 0.8 g agarose powder and made the final volume up to 100 ml with the agarose gel buffer in a flask. Heat the flask in a water bath and boil it for 30 min with occasional agitation.

## Procedure

1. **Prepare an agarose gel**: Take a flask containing 0.8% agarose gel and heat it in a microwave oven for 2~4 min until the agarose dissolves completely. Pour the warm agarose solution onto a glass slide, and insert a comb at the side 2 cm from cathode for forming the sample wells in the gel. Allow 20~30 min for solidification.
2. **Sample preparation**: Mix 0.2 ml serum with 0.02 ml dye solution and 0.01 ml ethanol. Settle the mixture at 37℃ for 30 min. Centrifuge it 5 min at 2000 r/min.
3. **Sample loading**: Carefully remove the comb and load 20 μl sample mixture into the wells. Place the gel in a horizontal electrophoresis apparatus. Connect the electrode and the gel with filter paper. Run the electrophoresis at a constant voltage of 120 V for 45~60 min.
4. Observe the shade of color and width of the bands by naked eyes to estimate their quantity, or estimate their percentages by scanning with densitometry.

## Reference

Reference value ranges of fasting serum lipoproteins in normal adults:
α-lipoprotein : (25.7 ± 4.1) % ;
pre-β-lipoprotein : (21 ± 4.4) % ;
β-lipoprotein : (53.3 ± 5.3) % .

## Clinical significance

Electrophoretic analysis of serum lipoproteins has great clinical significance. It can be used to detect primary disorders of plasma lipoproteins such as hyperlipoproteinemia, which are the metabolic disorders that the concentrations of specific lipoproteins in fasting serum were abnormally elevated. It can be classified into five types:

Type I : CM ( + )
Type IIa : β lipoprotein ( LDL ) ↑
Type IIb : β lipoprotein ( LDL ) ↑ , pre-β lipoprotein ( VLDL ) ↑
Type III : β lipoprotein ( LDL ) ↑ ↑
Type IV : pre-β lipoprotein ( VLDL ) ↑
Type V : pre-β lipoprotein ( VLDL ) ↑ , CM ( + )

# Chapter II  Electrophoresis

**Experiment title**

**Date**

**Observations and results**

## Discussion

1. Agarose gel has a macroreticular structure, and can be obtained in a form with zero ____.
2. Serum lipoproteins can be separated into at least four kinds of bands by agarose gel electrophoresis, including _____, _____, _____ and _____. They are related to the different lipoprotein layers obtained by ultracentrifugation _____, _____, _____ and _____.
3. Which bands may disappear in the electrophoresis of lipoprotein in the fasting serum of a healthy subject?

**Teacher's remarks**

**Signature**

**Date**

Experimental Manual in Medical Biochemistry

## Experiment 5  Proteins Separation by SDS-Polyacrylamide Gel Electrophoresis

### Principle

Polyacrylamide gel electrophoresis (PAGE) is a technique that uses a gel made of polymerized acrylamide. Monomers of acrylamide, $CH_2 = CH-C(=O)-NH_2$, are polymerized into long linear polymer, polyacrylamide, which can be cross-linked with N, N'-methylene bisacrylamide, $CH_2 = CH-C(=O)-NH-CH_2-NH-C(=O)-CH = CH_2$, to form a gel matrix. Polymerization is catalyzed by free radicals, generated by agents such as ammonium persulfate (AP) in the presence of N, N, N', N'-tetramethylethylenediamine (TEMED). The gel pore size is determined by the total acrylamide concentration and degree of cross-linking that regulated by varying ratios of acrylamide to bis-acrylamide. If the gel pore size approximately matches the size of the molecules to be separated, smaller molecules can move freely in an electric field, whereas larger molecules have restricted movement. Thus, proteins separate into discrete bands based on the gel pore size and protein's charge, size and shape.

A discontinuous gel system is also used frequently in PAGE, consisting of a stacking gel and a separating gel. The stacking gel makes up only approx 10% of the total gel volume, is of a lower percentage acrylamide (usually 2.5% ~ 5%) and a lower pH (6.8), compared with the separating gel, usually containing a higher percentage acrylamide (7% ~ 15%) and at a higher pH (8.8). The lower pH in stacking gel develops a sharp interface between the high mobile $Cl^-$, and the less mobile glycine$^-$. As the interface moves downward, the protein molecules with mobilities intermediate between $Cl^-$ and glycine$^-$, will be concentrated into a very thin band. When protein molecules reach the junction between a large pore stacking gel and a smaller pore running gel, they are more concentrated.

Sodium dodecyl sulfate (SDS)-PAGE, was introduced in 1967 by Shapiro *et al.* and become one of the most popular PAGE of proteins. SDS is an anionic detergent having a 12 carbon hydrophobic chain and the negative charged sulfonic acid group. The proteins interacting with SDS are given rod-like shape, because the detergent disrupts native ionic and hydrophobic interactions. SDS coats proteins with an approx. uniform charge-to-mass ratio (approx. 1.4 g SDS/g protein). In addition, disulfide bonds of proteins between Cysteine residues can be cleaved by a reducing agent such as β-mercaptoethanol or 1, 4-dithiothreitol (DTT). Thus, in denatured and reduced condition, separation of proteins or polypeptides is due solely to differences in size, and this method can be used to determine molecular weight (MW). That is say, a calibration curve can be generated by plotting relative mobilities of standard proteins (i.e. mobility relative to bromophenol blue tracker dye) versus their log MW, and subsequently used to estimate the MW of an unknown protein running on the same gel (Figure 2-6). The estimation of MW by this method is accurate to approximately 5% ~ 10% of the actual value.

Some proteins having color can be seen directly on a gel, but most proteins in the gel are typically visualized as blue bands by staining with Coomassie brilliant blue (CBB), which is one of the simplest non-radioactive methods. It has a reported sensitivity of 0.1 μg protein/band.

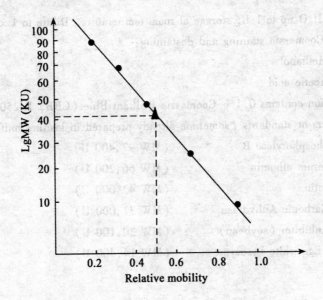

Fig. 2-6  Standard curve for protein MWs by SDS-PAGE

## Reagents

1. 30% Acrylamide stock solution: 29 g acrylamide and 1.0 g bisacrylamide dissolve in 100 ml dH$_2$O and through 0.45 μm filter. Stable at 4℃ for months. (caution: it is a neurotoxic).
2. 1.5 mol/L Tris-Cl buffer, pH 8.8.
3. 0.5 mol/L Tris-Cl buffer, pH 6.8.
4. 10% (W/V) Sodium dodecyl sulfate (SDS).
5. 10% (W/V) Ammonium persulfate (AP) (store at 4℃ 1~2 weeks).
6. 10% TEMED.

7. distilled water (dH$_2$O).
8. 2 × Sample loading buffer:
   - 100 mmol/L Tris-Cl, pH 6.8
   - 100 mmol/L DTT
   - 20% glycerol
   - 4% SDS
   - 0.2% Bromphenol blue (BB)
9. 5 × SDS-PAGE running buffer, pH 8.3

   | | |
   |---|---|
   | Tris Base | 15.1 g |
   | Glycine | 72.0 g |
   | SDS | 5.0 g |

   Dissolve in dH$_2$O up to 1 L, storage at room temperature. Dilute to 1 × before use.
10. Solutions for Coomassie staining and destaining:
    - 40%  Methanol
    - 10%  Acetic acid

    Staining solution contains 0.1% Coomassie Brilliant Blue (CBB) R-250.
11. Protein markers or standards (sometime already prepared in loading buffer):

    | | |
    |---|---|
    | Rabbit phosphorylase B | (MW 97,400 U) |
    | Bovine serum albumin | (MW 66,200 U) |
    | Rabbit actin | (MW 43,000 U) |
    | Bovine Carbonic Anhydrase | (MW 31,000 U) |
    | Trypsin Inhibitor (soybean) | (MW 20,100 U) |
    | Chicken Egg white Lysozyme | (MW 14,400 U) |

## Procedure

### 1. Preparation of the gel

(1) Gel Cassette Assembly: Clean and completely dry the glass plates, combs, and any other pertinent materials. Assemble the glass plate sandwich of the electrophoresis apparatus according to manufacturer's instruction: place a short plate on top of a spacer plate, insert both plates into the casting frame on a flat surface and clamp.

(2) Pour the Separating Gel: Prepare the separating gel solution for a vertical slab gel with size of 1 mm × 10 cm × 8 cm as directed in table 2-2. Add 10% AP and 10% TEMED just prior to pouring gel, stir gently to mix, then immediately apply to the sandwich along an edge of the spacers using a Pasteur pipet until the gel height is lower 1.5 cm than the short plate. Allow the gel to polymerize 30 ~ 60 min by covering gently with a layer of water (1 cm).

(3) A sharp optical discontinuity at the overlay/gel interface is visible on polymerization, then

pour off the water layer.

(4) Pour the Stacking Gel: Prepare the stacking gel solution as directed in table 2-2, adding 10% AP and TEMED, and pour. Insert the comb and allow the gel to polymerize completely.

Table 2-2　　　Recipes for polyacrylamide separating and stacking gels

| Reagents | 12% Separation Gel (ml) | 5% Stacking Gel (ml) |
|---|---|---|
| Distilled $H_2O$ | 4.8 | 2.72 |
| 1.5 mol/L Tris-Cl, pH 8.8 | 3.8 | - |
| 0.5 mol/L Tris-Cl, pH 6.8 | - | 1.25 |
| 30% Acrylamide stock solution | 6.0 | 0.83 |
| 10% SDS | 0.15 | 0.05 |
| 10% AP | 0.15 | 0.05 |
| 10% TEMED | 0.1 | 0.1 |
| Total volume | 15 | 5 |

## 2. Sample preparation
(1) Dilute a portion of the protein sample with 2 × SDS sample loading buffer in Eppendorf tubes (including preparation of protein markers.).
(2) Heat all samples for 3 ~ 5 min at 100℃ (a boiling water bath) to fully denature the proteins. Briefly centrifuge tube to accumulate protein samples at the bottom of tube.

## 3. Running the gel
(1) Carefully remove the comb, and then remove the gel cassette from the casting stand, and place it in the electrode assembly with the short plate on the inside. Place buffer dam plate opposite the gel cassette assembly. Provide a slight pressure on the gel cassette and buffer dam while clamping the frame to secure the electrode assembly.
(2) Place the assembly into the electrophoresis tank and completely fill the inner chamber with 1 × running buffer. Check for leaks. Use a gel-loading tip or syringe to pipette buffer into each well to remove bubbles.
(3) Slowly pipette 10 ~ 20 μl of denatured sample or protein markers at the bottom of the individual wells. (25 ~ 50 μg of total protein is recommended for a complex mixture.)
(4) Add enough running buffer to the region outside of the frame to cover platinum electrode at the bottom. Cover the tank with the lid aligning the electrodes (black or red) appropriately. Connect the electrophoresis tank to the power supply.
(5) Run the gel at 10 mA of constant current until BB dye enters the separating gel, then

increase the current to 20 ~ 30 mA and run until BB dye reaches the bottom of the gel. This can take as long as 1 hour and depends on the gel composition.

(6) When electrophoresis is complete, turn off the power supply, disassemble the apparatus, a gel knife is used to open the plates for exposing the gel.

## 4. Proteins detection by staining

(1) Remove the gel into a container and cover with CBB staining solution. Agitate on a gently orbital shaker for at least 20 min. Increase the time to improve detection of possible faint protein bands.

(2) Recycle the CBB staining solution and cover the gel with destaining solution, allow it to agitate slowly, carefully pour off destaining solution and replace with fresh until blue bands and a clear background are obtained.

(3) Destained gels can be rinsed stored in $dH_2O$ or dried by a gel dryer.

## Notes

(1) The purpose of glycerol in sample loading buffer is to increase the solution density so that the sample will sink beneath the running buffer and stay in the sample loading well, and adding BB (a tracker dye) is to indicate how the electrophoresis is progressing because it migrates ahead of proteins in the gel.

(2) Because polymerization of polyacrylamide is inhibited by oxygen, so it must be cast into a sealed mould. Covering with a layer of water can isolate the separating gel from oxygen and form a smoothing gel surface.

(3) unpolymerized acrylamide and bis acrylamide are neurotoxic. When preparing and handling gels, use gloves to avoid absorbed through the skin.

Chapter II  Electrophoresis

**Experiment title**
**Date**
**Observations and results**
(Measure the migration distance of each protein, and construct a calibration curve. Calculate the MW of the unknown protein according to the curve)

**Discussion**

1. Acrylamide stock solutions for PAGE are typically made up of _____ and _____, which to create pores of different size. Gel Polymerization is catalyzed by free radicals, generated by agents such as _____ in the presence of _____.
2. A discontinuous gel system is used frequently to _____ samples, it consists of a _____ gel and a _____ gel.
3. In PAGE, the combination of gel pore size and protein _____, _____, and sharp determines the migration of the protein. In SDS-PAGE, separation of proteins is solely on the basic of their _____.
4. What was the role of SDS and DTT in the gel loading buffer?

**Teacher's remarks**
**Signature**
**Date**

# Chapter III

# Chromatography

In 1903, Russian scientist Mikhail Tswett invented a technique to separate different plant pigments from green leaves. This method was named chromatography (Chroma means color and graphy means atlas). Chromatography is the laboratory technique that allows separation of different components in the mixture based on their differences of physical and chemical properties including absorption capacity, solubility, molecular shape and size, molecular polarity and binding affinity for a particular ligand or a solid support. Nowadays, chromatography has been developed into a multidisciplinary technology. Due to its high efficiency, high sensitivity and high resolution potency, chromatography is widely applicable in industry, agriculture, biochemistry, chemical engineering, medicine science, and so on.

## 3.1 Fundamental Principles of Chromatography

### 3.1.1 General principles of chromatography

Although there are many different types of chromatography, the principle of them for separating of the molecules are almost the same. Generally, there are two phases for a chromatography: stationary phase and mobile phase. In the process of chromatography, stationary phase is kept stationary state, while the mobile phase flows over or through the stationary phase. The phases are chosen so that components of the mixture have different distribution in each phase. Components distributing preferentially in the mobile phase will pass through the chromatographic system faster than those distributing preferentially in the stationary phase. Due to their differences of mobility, components can be differentially eluted from the stationary phase following with the mobile phase, and then be detected and alalyzed. This process of eluting is named as "development".

### 3.1.2 Classification of chromatography

There are many classification schemes for chromatography.

(1) According to the physical nature of the mobile phase, chromatography can be divided into liquid chromatography (LC) and gas chromatography (GC). The stationary phase can also take two forms, solid and liquid, which provides subgroups of LC and GC: liquid-solid chromatography (LSC), liquid-liquid chromatography (LLC), gas-solid chromatography (GSC) and gas-liquid chromatography (GLC).

(2) According to the distribution style, chromatography can be divided into adsorption chromatography, partition chromatography, ion exchange chromatography, gel chromatography and affinity chromatography according to the principle of chromatography.

(3) According to the mode of the stationary phase, chromatography can be divided into column chromatography and plane chromatography according to the modes of separation. Plane chromatography includes paper chromatography, thin layer chromatography, thin film chromatography and so on. Modern chromatography is often performed in a column format. Here, column chromatography will be discussed in details.

### 3.1.3 Basic components of column chromatography

Column chromatography consists of a solid stationary phase and a liquid mobile phase. The stationary phase is confined to a column and the mobile phase (a buffer or solvent) is allowed

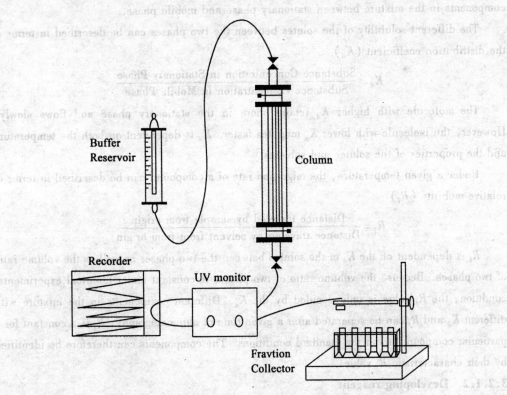

Fig. 3-1 Basic components of column chromatography

to flow through the solid phase in the column. A basic apparatus for column chromatography systems mainly comprise a buffer reservoir, a chromatographic column, a detector, a recorder and a fraction collector. The buffer reservoir contains liquid mobile phase delivered either by gravity or using a pump. The chromatographic column is the core part of chromatography separation. Generally, it is a glass or plastic column containing stationary phase. The solution flows out from the chromatography column to the automatic fraction collector through a detector, such as the ultraviolet monitor. The signal detected by the detector is recorded by the recorder. The outcome of a chromatography experiment can be shown as a chromatogram. The complete chromatographic system may also contain an autosampler to add samples and a computer to control the system.

## 3.2 Commonly Used Chromatographic Techniques

### 3.2.1 Partition chromatography

#### 3.2.1.1 Principle

Partition chromatography is a separation technique based on the different solubility of the components in the mixture between stationary phase and mobile phase.

The different solubility of the solutes between the two phases can be described in terms of the distribution coefficient ($K_d$)

$$K_d = \frac{\text{Substance Concentration in Stationary Phase}}{\text{Substance Concentration in Mobile Phase}}$$

The molecule with higher $K_d$ retains more in the stationary phase and flows slowly. However, the molecule with lower $K_d$ migrates faster. $K_d$ is dependent on both the temperature and the properties of the solutes and solvents.

Under a given temperature, the migration rate of a compound can be described in terms of relative mobility ($R_f$)

$$R_f = \frac{\text{Distance traveled by sample from origin}}{\text{Distance traveled by solvent front from origin}}$$

$R_f$ is dependent on the $K_d$ of the solutes between the two phases as well as the volume ratio of two phases. Because the volume ratio of two phases is constant under identical experimental condition, the $R_f$ value is only decided by the $K_d$. Different components in the mixture with different $K_d$ and $R_f$ can be separated after a given time of chromatography. $R_f$ is a constant for a particular compound under the standard conditions. The components can therefore be identified by their characteristic $R_f$ values.

#### 3.2.1.2 Developing reagent

In normal-phase partition chromatography, the stationary phase is polar solvent (such as water bounded to the filter paper in paper chromatography), and the mobile phase used in development is nonpolar solvent (usually combinations of organic and aqueous solvents). In reverse-phase partition chromatography, the stationary phase is nonpolar and the mobile phase is polar.

#### 3.2.1.3 Detection and analysis of components

The colored sample bands or sample spots can be observed during the process of chromatography. But if the sample is colorless, the sample bands or spots have to be colored by spraying reagent for reaction. The $R_f$ value of different components in a sample are calculated, and components can be identified based on their $R_f$ values as well as those of the standards.

### 3.2.2 Ion exchange chromatography

#### 3.2.2.1 Principle

Ion exchange chromatography is a separation method based on charge differences. There are two types of ion exchange chromatography, namely cation exchange chromatography and anion exchange chromatography. The stationary phases are ion exchangers which are either positively or negatively charged on their surface. The mobile phases are electrolyte solutions with a particular pH and ionic strength. The choice of which type of ion exchanger, as well as the composition of the buffers used in the experiment should be determined in advance.

Take cation exchange chromatography as an example: cation exchange refers to a situation in which the stationary phase carries negative charges and has affinity for cations in solution. When a mixture of charged molecules passes through the column containing the cation exchanger, the molecules with strong positive charges will bind much tightly than those carrying weak positive charges. The bound molecules can be eluted out one by one by an eluting buffer with a suitable pH and ion strength or a pH gradient buffer.

Ion exchange chromatography can be used to separate mixtures of biological molecules with ionized groups including amino acids, polypeptides, proteins, and nucleotides.

#### 3.2.2.2 Ionic exchanger

The basic requirements of ion exchanger are loose and porous structure or huge surface area, allowing free diffusion and exchange of charged molecules in it, highly insolubility, more exchangeable groups, physical and chemical stability

Depending on their composition and properties, the ion exchangers can be classified as hydrophobic and hydrophilic ion exchanger. The hydrophobic ion exchangers are made by co-polymerizing styrene with divinylbenzene, and are normally used to separate small molecules. The hydrophilic ion exchangers including cross-linking agarose, cross-linking sephadex, cellulose and polyacrylamide with varying degrees of cross-linking, are normally used to separate biomacromolecules such as nucleic acids and proteins.

Depending on the exchangeable ions and their exchange properties, the ion exchangers can be classified as cation exchanger and anion exchanger. The commonly used cation exchangers are polystyrene sulfonic acid cation exchange resin, carboxymethyl cellulose (CM-cellulose) and CM-Sephadex C-50. The commonly used anion exchangers are diethylaminoethyl cellulose (DEAE-cellulose) and DEAE-Sephadex A-50.

### 3.2.3 Size exclusion chromatography/ Gel filtration

#### 3.2.3.1 Principle

Size exclusion chromatography, also named as gel filtration, is a separation method based on the differences of molecular weight of components in the mixture. The stationary phases are gel granules that have the effect of molecular sieve (porous matrix). When a solution containing components with different molecular weight passes through the column containing a porous gel matrix, large molecules cannot enter into the pores of gel granules and will be excluded rapidly. Smaller molecules will easily enter the pores of porous gel particles and therefore be eluted out at a slower rate. Figure 3-2 illustrates the principle of gel filtration.

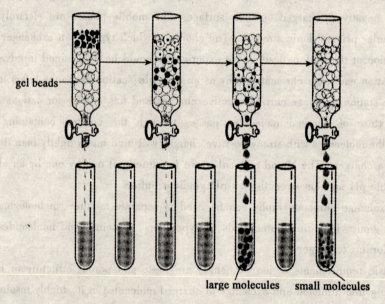

Fig3-2  The principle of gel filtration

To assess the extent to which sample molecules can penetrate the pores in the stationary phase, the distribution coefficient ($K_d$) of a particular solute between the inner and outer solvent are routinely measured. For a given type of gel, $K_d$ is a dependent on molecular size of the solute and is given by:

$$K_d = \frac{V_e - V_o}{V_i} = \frac{V_e - V_o}{V_t - V_o}$$

Where $V_e$ (elution volume) is the volume of the mobile phase that is required to elute a particular compound, $V_o$ (void volume) is the volume of mobile phase between the beads of the stationary phase inside the column, $V_i$ (inner volume) is the volume of mobile phase inside the porous beads. $V_t$ is total bed volume, $V_t = V_o + V_i + V_g$ ($V_g$ is the volume occupied by matrix, it is often neglected because it is only a small part of $V_t$). If the compounds are too large to fit inside any of the pores of gel beads and completely excluded from the solvents within the gel, $K_d = 0$, and they will eluted at first. If the compounds are small enough to fit inside all the pores of the gel beads, $K_d = 1$, and they will be last eluted. Compounds of intermediate size are partially fit inside some but not all of the pores of gel beads. These compounds will be eluted between the large ("excluded") and small ("totally included") ones, and their $K_d$ values be ranged between 0 and 1.

### 3.2.3.2 Gel matrix

The gel matrix for gel filtration consists of porous beads with a well-defined range of pore sizes. The commonly used gels include cross-linked dextran polymer (trade mark Sephadex, or Sephacryl), cross-linked agarose (trade mark Sepharose), cross-linked polyacrylamide (trade mark Biogel) and so on.

The dextran gels are porous gel granules cross linked with polysaccharide dextran by chemical method. The gel granules with different degrees of cross linking have different pore sizes. According to the degree of cross-linking, Sephadex may be divided into different types that can be indicated by "water regain", i.e., the amount of water taken up in the completely swollen gel granules per gram of dry gel. For example, if the water regain by one gram of Sephadex is 25 grams, it will be named as Sephadex G-25.

Polyacrylamide gels are prepared by the polymerization of acrylamide and methylene bisacrylamide. By varying the relative proportions of the two monomers, a range of gels with different pore sizes can be obtained. Take Biogel P-6 as an example, it has a molecular weight exclusion limit of 6,000 U.

Agarose gels are porous gel granules cross linked by different concentrations of agarose. The commonly used types are Sepharose 2B, 4B and 6B, which represents 2%, 4%, 6% of agarose respectively. This type of gel matrix has satisfactory mechanical property, high resolution and broad separation scope. Like dextran gels, agarose gels cause very little denaturation and absorption of sensitive biological molecules and therefore are widely used in separation of biomacromolecules.

The commercial gel types suitable to differentiate the molecules with the relative range of size are listed in table 3-1.

Table 3-1　　　　　　　　　Commonly used gel matrix for gel filtration

| Gel Matrix | Fractionation Range (U) |
|---|---|
| Sephadex G-15 | 50 ~ 1 000 |
| Biogel P- | 50 ~ 1 000 |
| Sephadex G-25 | 1 000 ~ 5 000 |
| Sephadex G-50 | 1 500 ~ 30 000 |
| Biogel P-10 | 1 500 ~ 30 000 |
| Sephadex G-100 | 4 000 ~ 150 000 |
| Sephadex G-200 | 5 000 ~ 250 000 |
| Sephacryl S300 | 20 000 ~ 1 500 000 |
| Sepharose 4B | 60 000 ~ 20 000 000 |

#### 3.2.3.3　Application of gel filtration

Gel filtration can be used to separate mixtures of biomacromolecule, especially enzymes, antibodies and some other globular proteins. It can also be used for de-salting, concentration of macromolecular solution, determination of molecular mass of biomacromolecules and removing pyrogenic substances.

### 3.2.4　Affinity chromatography

#### 3.2.4.1　Principle

Affinity chromatography is a separation method based on the reversible binding of a biomacromolecule to a complementary binding substance (ligand). There is a specific binding capacity between enzyme and substrate analogue, antigen and antibody, avidin and biotin, hormone and its receptors, RNA and its complementary DNA. Affinity chromatography exploits those biospecific relationships to specifically separate the target molecule from a mixture. The biospecific ligand, which is covalently attached to a matrix as the stationary phase, binds to the target molecule within a solution to be analyzed. Ideally, the ligand will bind the target molecule only and all other unbound compounds in the solution will be washed out. Target protein is recovered by changing conditions to favor elution of the bound molecules (typically by varying the pH or increasing the ionic strength or adding inhibitors). The principle of affinity chromatography is shown in figure 3-3. Affinity chromatography has been widely used in protein purification, nucleic acid purification, antibody purification from blood serum and separation of a mixture of cells into homogeneous populations.

#### 3.2.4.2　Ligand and matrix

A successful separation of affinity chromatography requires a unique biospecific ligand.

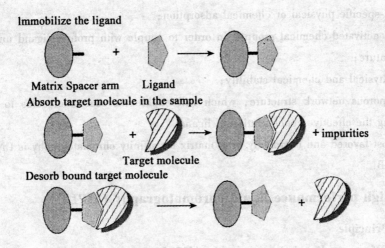

Fig. 3-3 The principle of affinity chromatography

Therefore, the appropriate selection of ligand is very important. Generally, it is selected according to its magnitude of affinity and specificity to the target molecule. The commonly used affinity systems are listed in table 3-2.

Table 3 – 2  Commonly used affinity system

| Ligand | Respective binding substance |
| --- | --- |
| enzyme | substrate analogue, inhibitor, cofactor |
| antibody | antigen, virus, cell |
| lectin | cell surface receptor, polysaccharides, glycoprotein |
| hormone | receptor, hormone carrier protein |
| biotin | avidin |
| nucleic acid | complementary base sequence, histone |
| $Ni^{2+}$ | protein with His 6 tag |
| Protein A | IgG |

The specific ligand is immobilized on an insoluble matrix, in a manner which does not interfere with its interaction between ligand and the respective binding substance. This may require the use of a spacer arm, which typically consists of a chain of about 6 ~ 10 carbon atoms to facilitate effective binding.

The ideal matrix used in chromatography should have following characteristics:

(1) insoluble in water but with high hydrophilicity;
(2) no non-specific physical or chemical adsorption;
(3) enough activated chemical groups, in order to couple with profuse ligand under optimum temperature;
(4) good physical and chemical stability;
(5) loose porous network structure, which will enable the macromolecules to pass freely, enabling the effective concentration of ligand.

The most favored and commonly used matrix for affinity chromatography is CNBr-activated Sepharose-4B.

### 3.2.5 High performance liquid chromatography (HPLC)

#### 3.2.5.1 Principle

High performance liquid chromatography (HPLC) has the same principle as other classical chromatography methods. Technically, the mobile phase is transported by high-pressure through the chromatography column, which is padded with particles (3 ~ 10 microns) of stationary phase by special method. Its efficiency is much higher than the general liquid chromatography. Simultaneously, a high sensitivity detector is connected to the column, which can examine the flowing liquid continuously. HPLC is divided into analysis mode and preparation mode according to the experimental purpose. The analysis mode is mainly used in the ingredient analysis of the sample, and the preparation mode is mainly used for the preparation of some substances. The stationary phase in HPLC refers to the solid gel matrix and the mobile phase refers to the solvent acting as a carrier for the sample solution. Different stationary phases and mobile phases are selected according to the experimental purpose.

Generally, the smaller the particle size of the stationary phase is, the greater the resolving power of chromatographic column is. However, the particle with smaller size will cause the greater resistance to eluant flow, and will elongate elution time for samples. In order to reduce the elution time, the high pressure pumping systems are needed for the supply of eluant to the column. However, high pressure is not an inevitable pre-requisite for high performance. New smaller particle size stationary phases that can withstand high pressure are available now.

#### 3.2.5.2 Apparatus of a HPLC system and its applications

A HPLC system mainly contains gradient mixers, sampling system with accurate sample valves, high pressure pumps with very constant flow, high efficiency HPLC columns, high sensitive detection system and the computer automatic data processing system.

HPLC has emerged as the most popular, powerful and versatile form of chromatography. HPLC has been developed for many of the types of chromatography including normal-and reverse-phase partition, ion exchange, gel filtration and so on. Preparative HPLC can be used for the separation and purification of proteins, peptides, and polynucleotides. The analytical application of HPLC includes quality control, process control, forensic analysis, environmental monitoring and clinical testing.

Experimental Manual in Medical Biochemistry

# Experiment 6  Detecting Transamination of Amino Acids by Paper Chromatography

## Principle

Transamination refers to the transferring of an amino group from an α-amino acid to an α-keto acid, which is catalyzed by transaminase (or aminotransferase). There are specific transaminases for most amino acids. One of the most important transaminases is glutamate-pyruvate transaminase (GPT), also known as alanine transaminase (ALT). It is a hepatocellular enzyme that catalyzes the following reaction:

$$\text{Alanine} + \alpha-\text{ketoglutarate} \xrightarrow{\text{ALT}} \text{Pyruvate} + \text{Glutamic acid}$$

In this experiment, α-ketoglutaric acid and alanine are used as substrates. After incubation with liver homogenate, the formation of glutamic acid by transamination is demonstrated by paper chromatography.

Paper chromatography is a type of partition chromatography. In paper chromatography, the mobile phase is a solvent mixture of water and organic solvents (developing solvent). As the solvent mixture moves up the paper by capillary action, the water in the mixture binds to the hydrophilic filter paper and creates a stationary phase. A polar amino acid will dissolve more in the stationary water phase and will move up the paper slightly. A non-polar amino acid will be more soluble in the mobile non-polar solvent than in water, so it will continue to move up the paper. The amino acids have different distribution coefficients between the stationary and mobile phases, thus they have different relative mobility ($R_f$) and then can be separated and identified.

## Reagents

1. 0.01 mol/L Sodium phosphate buffer (PBS, pH 7.4): Mix 81 ml 0.2 mol/L $Na_2HPO_4$ and 19 ml 0.2 mol/L $NaH_2PO_4$, dilute 20 fold with distilled water.
2. 0.1 mol/L Alanine: Dissolve 0.891 g alanine in a suitable amount of 0.01 mol/L PBS, adjust the pH to 7.4 with 1 mol/L NaOH, and bring to 100 ml with 0.01 mol/L PBS.
3. 0.1 mol/L α-ketoglutarate: Dissolve 1.461 g α-ketoglutarate in a suitable amount of 0.01 mol/L PBS (pH 7.4), adjust the pH to 7.4 with 1 mol/L NaOH, and bring to 100 ml with

0.01 mol/L PBS.
4. 0.1 mol/L Glutamic acid: Dissolve 0.735 g glutamic acid in a suitable amount of 0.01 mol/L phosphate buffer (pH 7.4), adjust the pH to 7.4 with 1 mol/L NaOH, and bring to 50 ml with 0.01 mol/L PBS.
5. 0.5% Ninhydrin: Dissolve 0.5 g ninhydrin in 100 ml acetone.
6. Developing solvent: Phenol: $H_2O = 4:1(V/V)$, prepared fresh.

## Procedures

1. Preparation of liver homogenate
   Homogenize 1 g fresh mouse liver in 2 ml 0.01 mol/L PBS (pH 7.4) on ice.
2. Transaminaton reactions
   Two test tubes are used and the operation is done according to the following table.

| Reagents (ml) | Transamination tube | Blank tube |
|---|---|---|
| Liver homogenates | 0.5 | 0.5 |
| Bath the blank tube in boiling water for 5 min and cool. | | |
| 0.1 mol/L alanine | 0.5 | 0.5 |
| 0.1 mol/L α-ketoglutaric acid | 0.5 | 0.5 |
| 0.01 mol/L phosphate buffer | 1.5 | 1.5 |

Mix the contents of tubes, incubate 60 min at 37℃. Incubate the transamination tube in boiling water bath for 5 min to terminate the reaction. After cooling the tubes, centrifuge the two tubes at 2000 r/min for 5 min. Transfer the supernatant into two new tubes labeled correspondingly.

3. Paper Chromatography
(1) Preparative step. During manipulation that follows, minimize the contact of paper with your hands to avoid contamination. Take a piece of round-shaped filter paper, and lay it on a sheet of white clean paper. Draw with a sharp pencil a small circle at the center with a diameter of 0.7 cm as the base line. Divide the circle into four equal parts and mark the four points as indicated in Figure 3-4 as the origins.
(2) Sample application. The samples are applied to the origins by capillary tubes. Dip the end of capillary tube into the sample solution, the solution will go into the capillary tube spontaneously by capillary action. Touch the capillary tube end to the origins on the filter paper, a sample spot will be formed. The diameter should be less than 4 mm, and usually the smaller, the better. If the sample solution is too diluted, the sample application process may be repeated 1 ~ 2 times. Be sure that the previous spot is completely dried

before you apply the sample for the second time at the same site. Apply the solution transamination tube to the origin point 1, the solution in blank tube to the origin point 3, the standard alanine solution to origin point 2, and the standard glutamic acid solution to origin point 4. Prevent cross pollution when apply solution with capillary tubes.

(3) Development. Make a small hole on the center of the round filter paper. Then take a filer paper strip, roll it into a wick, insert the upper end into the center hole of the filter paper. Add about 1 ml of developing solvent (phenol: $H_2O$ = 4:1) onto a watch glass placed in a Petri dish. Place the filter paper on the upper edge of the Petri dish with the filer paper wick dipped into the developing solvent. Cover the filter paper with another Petri dish. The developing solvent runs through the wick to the filter paper and diffuses in a circle (chromatography time is approximately 45 ~ 60 min). When the solvent front nearly reaches the edge of the filter paper, stop developing by removing the wick. Mark the solvent front quickly with pencil, and dry it by blowing hot air with a hair drier.

(4) Visualization. After drying, the filter paper is sprayed with 0.5% ninhydrin solution. The sprayer may be blown by a blower. Take care not to spray too much ninhydrin solution. Record the color intensity of each patch. Measure the distance from the origin to the solvent front and the distance from the origin to the center of each patch. Calculate the $R_f$ values of each patch using the following formula. Explain the transamination reactions occurred based on the data.

$$R_f = \frac{\text{Distance traveled by sample from origin}}{\text{Distance traveled by solvent front from origin}}$$

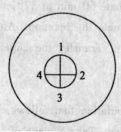

Fig. 3-4  Sample application points

# Chapter III  Chromatography

**Experiment title**

**Date**

**Observations and results**

## Discussion

1. Briefly explain your results of the experiment. If the control tube has not been incubated in boiling water enough, what might be the result?

**Teacher's remarks**

**Signature**

**Date**

Experimental Manual in Medical Biochemistry

| Experiment 7 | Separation of Mixed Amino Acids by Cation Exchange Chromatography |

## Principle

Ion exchange chromatography is a technique to separate mixtures of charged compounds. It relies on the differential electrostatic affinities of charged molecules for a charged stationary phase (ion-exchange resin). Amino acids are ampholytes. The net charge on an amino acid depends on their pI and on the pH of the solution. Different amino acids will carry different charges in a solution with a given pH. The differently charged amino acids can be separated by being eluted from ion exchange columns according to their different binding abilities to the ion exchanger. This experiment uses sulfonic acid cation exchange resin to separate the mixture of acidic amino acid (Aspartic acid, Asp, pI = 2.97) and basic amino acid (Lysine, Lys, pI = 9.74). In the condition of pH = 5.3, lysines are positively charged and can bind to cation exchange resin. Aspartic acids are negatively charged and can be washed out without any binding. In the condition of pH = 12, lysines are negatively charged and may then be eluted out. So they can be respectively eluted and separated by changing the pH or the ion strength of eluting solution.

## Reagents

1. Resin: sulfonic acid cation exchange resin (732 type).
2. Eluting solution:
(1) 0.45 mol/L Citric acid buffer (pH 5.3): dissolve 285 g citric acid and 186 g sodium hydroxide in distilled water. Adjust the pH to 5.3 with about 105 ml hydrogen chloride. Bring the total volume of the solution to 10 L with distilled water ($dH_2O$).
(2) 0.01 mol/L Sodium hydroxide (pH 12): dissolve 4 g sodium hydroxide in 10 L of $dH_2O$.
3. Mixture of amino acids: 5 mmol/L Asp and Lys in 0.02 mol/L hydrogen chloride.
4. Dye reagent (titanium trichloride-ninhydrin solution).
(1) Acetic acid buffer (pH 5.5): dissolve 82 g sodium acetate in 100 ml $dH_2O$. Then add 25 ml acetic acid. Dilute to 250 ml with $dH_2O$.
(2) Dissolve 20 g ninhydrin in 750 ml ethylene glycol monomethylether, add 1.7 ml titanium trichloride, then make the final volume 1 L with acetic acid buffer.

## Procedure

1. The treatment of resin

   Put about 10 g of resin in a 100 ml beaker and swell it thoroughly with water. Discard the water. Add 25 ml of 2 mol/L hydrogen chloride, and stir for 30 min. Discard the acid liquor, and wash the resin completely to neutral with $dH_2O$. Add 25 ml of 2 mol/L sodium hydroxide to the resin and stir for 30 min. Discard the basic liquor, and wash the resin to neutral with $dH_2O$. Suspense the resin in 100ml citric acid buffer (pH 5.3).

2. Packing

   Take a chromatography column with the diameter of 0.8 ~ 1.2 cm and the length of 10 ~ 12 cm, put a piece of glass cotton circular cushion at the bottom, and pour the above prepared resin from the top. Close the exit of chromatography column. When the resin sediment appears, let out the excessive solution, add more resin until the height of the resin sediment is about 8 ~ 10 cm.

3. Equilibrium

   Add pH 5.5 citric buffer from the top to wash it until the flowing liquor reaches pH 5.5. Close the exit of the column. It is needed about 3 ~ 5 times column volume of citric acid buffer. Keep the level of the liquor surface about 1cm higher than that of the resin.

4. Loading and elution

   Open the exit to make the buffer flow out. When the level of the liquor is at almost the same height as that of resin, close the exit (Do not allow the resin to dry out!). Use a long pipette dropper to add 0.5 ml mixture of amino acid to the top of resin directly and carefully, and open the exit to make it flow into the resin slowly. Then wash the column wall twice by adding 0.5 ml citric acid buffer. Add citric acid buffer to the top of resin until the level of the liquor is 2 ~ 3 cm above that of resin. Then elute the resin at the flowing rate of 0.5 ml/min. Collect the fractions of 3.0 ml in each tube. After collecting 5 tubes, elute the resin with 0.01 mol/L sodium hydroxide (pH 12) and collect another 7 tubes.

5. Regeneration

   Wash the column with citric acid buffer (pH 5.3), and the column can be used again.

6. Amino acid detection

   After collecting the eluent, number the tubes collected. Transfer 0.5 ml of eluent from each of the tubes to a clean tube respectively. Add 1ml of pH 5.3 citric acid buffer and 0.5 ml of dye reagent into the tube. Mix the tubes and incubate for 25 min in the boiling water bath. Cool the tubes with running water. Add 3 ml of 60% alcohol into the tube and mix. The solution showing royal blue means some amino acid has been eluted out. The degree of the color can show the concentration of amino acid, and it can be determined by measuring the

Experimental Manual in Medical Biochemistry

optical density of every tube of eluent (take water as blank and at the wavelength of 570 nm). Finally, take the optical density of every tube of eluent as ordinate, and the tube number of eluent as abscissa, then draw an elution curve.

Chapter III  Chromatography

**Experiment title**
**Date**
**Observations and results**

## Discussion

1. The stationary phases in ion exchange chromatography are _____, the mobile phases are _____.
2. Why can the mixed amino acids (aspartate and lysine) be eluted from cation exchange resin one by one?

**Teacher's remarks**
**Signature**
**Date**

# Experiment 8  Separation of Hemoglobin and Dinitrophenyl (DNP)-glutamate by Gel Filtration Chromatography

## Principle

Gel filtration, also named as size exclusion chromatography, separates proteins based on their molecular size. In the experiment, Sephadex G-50 was used to separate the mixture of hemoglobin (Hb, red, MW 64500U) and dinitrophenyl (DNP)-glutamate (yellow, MW 2000 ~ 12000 U). Because of the differences of molecular weight, the two components will be eluted out of the column at different rates. In the process of elution, two colored bands will be observed: The large hemoglobin molecules will be eluted out first as a red band, and the smaller DNP-glutamate molecules will flow out of the column later as a yellow band.

## Reagents

1. Sephadex G-50.
2. Mixture of hemoglobin and DNP-glutamate.
3. distilled water ($dH_2O$).

## Procedure

1. Gel preparation

   Add 4 g of Sephadex G-50 to 30 ml $dH_2O$ and expand it overnight at room temperature or 2 h in boiling. Remove the upper water and small particles.

2. Packing

   Take a chromatography column with the diameter of 1 cm and the length of 25 cm, put a piece of glass cotton circular cushion at the bottom, and pour about 4 cm of buffer into the column. Close the exit of chromatography column. Add the slurry of the prepared gel by gently pouring it down the side of the column. Open the exit when the height of the gel deposited is about 1/3 of the length of the column. Continue stacking until the final surface of the gel bed is 3 cm below the top of the column. Close the exit of the column. It is vital that the column be set up properly so that it does not dry out.

3. Loading

   Add the sample to the top of the column bed using a transfer pipette. The tip of the pipette is held against the wall of the column and moved in a circular motion so the sample slowly goes down the gel bed, being sure not to disturb the gel. Open the exit of the column to allow the sample to enter the column. When the level of the liquor is at almost the same height as that of gel, close the exit (do not allow the gel to dry out!). Wash the top of the gel twice 5 ~ 10 drops of $dH_2O$, to ensure all of samples running into the gel.

4. Elution

Add $dH_2O$ to the column and let it flow. You should be able to see the separation of the colored bands almost immediately. When the first colored band is about to elute from the column, collect the sample with a test tube. Collect the sample with a new tube when the second colored band elutes from the column. Close the outlet and put the end cap back on the bottom of the column when all of the colored bands have eluted from the column.

Experimental Manual in Medical Biochemistry

**Experiment title**
**Date**
**Observations and results**

## Discussion

1. How many different colored bands appear as the sample separated through the column? What color is eluted first? What proteins make up each colored band?

**Teacher's remarks**
**Signature**
**Date**

# Chapter IV
# Enzyme Analysis

Enzymes are highly effective and extremely specific biochemical catalysts produced by living organisms and accelerate the metabolic reactions under the chemically mild conditions in the cell. Enzymes are typically proteins and their catalytic activity depends on the precise conformational structure in the folded polypeptide chains. Even minor alterations in this structure may result in the loss of enzyme activity. The catalytic properties of an enzyme may also depend on the presence of non-peptide cofactors or coenzymes. It is of practical significance to study the properties of enzyme in medicine.

## 4.1 The Activity of Enzyme

### 4.1.1 Enzyme activity determination

The enzymes of cellular metabolism are located within the tissue cells and are present there at high concentrations. When tissues break down or cell membranes leak, the level of these enzymes in the plasma rises. Measurement of their quantity of plasma may provide the qualitative or quantitative indexes of tissue damage. The quantity of enzyme present in a sample is usually referred to enzyme activity rather than weight or volume of an enzyme. Enzyme activity is defined as the ability of an enzyme to catalyze a specific reaction.

#### 4.1.1.1 Enzyme activity and the velocity of enzymatic reaction

Normally, the velocity of enzyme-catalyzed reaction is proportional to the enzyme activity. The faster the reaction velocity is, the higher the enzyme activity is. Therefore measurement of enzyme activity (actually it is a measure of the quantity of an active enzyme present) is the detection of the enzyme catalytic reaction velocity, which can be measured as the consumption of substrate or the accumulation of product per unit time under special conditions. To measure reaction velocity, some property difference between substrate and product must be identified. The reaction velocities employ the initial velocity under the conditions that only traces of product accumulates, hence the velocity of the reverse reaction is negligible.

Experimental Manual in Medical Biochemistry

### 4.1.1.2 Methods for measurement of enzyme activity

Methods for measurement of enzyme activity, also called enzyme assays, can be classified into two groups according to their sampling method: continuous assays, where the assay gives a continuous reading of activity, and discontinuous assays, where the reaction is stopped after a given time and then the concentration of substrates / products determines.

The commonly used assays for measurement of enzyme activity are chemical assay, spectrophotometric assay, electrochemical assay, fluorimetric assay and radiometric assay. The methods selected for enzyme activity assay are dependent on the physical and chemical characters of substrates or products.

Usually, the applied substrates are in excess of the demand in a reaction and the amount of reduced substrate only accounts for a little portion of the total substrates. Therefore it is difficult to determine the consumption of substrates accurately. However, the product newly formed in the reaction can be precisely measured by proper sensitive methods. Common assays of enzyme activity are introduced as following.

**(1) Chemical assay**

Chemical method is often a discontinuous assay, in which the enzyme-catalyzed reaction is stopped by adding the enzyme denaturant to the reaction system. The enzyme-catalyzed reaction can also be stopped by heating that can lead to thermal denaturation and inactivity of the enzyme. The method is tedious and time consuming. Moreover, it is not easy to obtain precise result for some rapid reactions.

**(2) Spectrophotometric assay**

Spectrophotometric assay is the most popular method based on the different spectral absorbance between substrate and product at certain wavelength. The change in absorbance can be used as the basis of quantitative analysis in the assay according to the Lambert-Beer'law. If the wavelength is in the visible region, a change in the color of the assay system can actually be seen, and these are called colorimetric assays. Many assays are based on the interconversion of $NAD(P)^+$ and $NAD(P)H$. The reduced form, $NAD(P)H$ has the maximum absorbance at 340 nm, but the oxidized form, $NAD(P)^+$ has not this character. Therefore the reaction catalyzed by the dehydrogenases using $NAD(P)H$ as coenzyme can be measured by monitoring the absorbance change at 340 nm.

**(3) Fluorimetric assay**

Fluorometric assay is a method to measure the intensity of fluorescence emitted from substrate, product or coenzyme. For example, NADPH, the reduced form of coenzyme of some dehydrogenases, can emit strong blue-white fluorescence in neural solution while the oxidized form ($NADP^+$) doesn't emit fluorescence. Many enzymes can be assayed by coupling with an appropriate reaction that involves NADPH. The enzyme activity can be expressed as the change of fluorescence intensity per unit time.

The advantage of fluorometric assay is its high sensitivity which is higher than that of spectrophotometric assay for 2 ~ 3 orders of magnitude. Thus it is extremely suitable for the rapid enzymatic analysis of enzymes or substances with a very low concentration. The disadvantage of fluorometric assays is that the fluorometric reading is not directly proportional to the concentration of substance. Furthermore, the results may vary with the experimental conditions such as temperature, scattering of light, equipment, etc.

(4) **Radiometric assay**

Radiometric assay is a method to measure the isotope incorporated into products or isotope release from substrates. The radioactivity can be measured after separating the substrates from the products, which can represent the increase of products or the decrease of substrates in the reaction. The radioactive isotopes most frequently used in this assays are $^{14}C$, $^{32}P$, $^{35}S$ and $^{125}I$. Radioactive isotopes assay is extremely sensitive and specific. However, this assay is extremely tedious, time-consuming and dangerous due to its radioactivity. Therefore the operator must be very cautious and the experiment should be performed in the special laboratory with the specific permintion for isotope usage.

### 4.1.2 The unit of enzyme activity and specific activity

In 1976, the International Enzyme Commission defined that one unit of enzyme activity is the amount of enzyme which will catalyze the transformation of 1 micromole of the substrate per minute under standard conditions. This unit has a symbol "U". Another unit of enzyme activity is the katal, which is the amount of enzyme that converts 1 mole of substrate per second. So 1 U = 1/60 micro katal = 16.67 nano katal (or 1 katal = $60 \times 10^6$ U).

In clinical lab, the unit of an enzyme activity is usually expressed in a traditional way as the detection method. For example, according to the King-Armstrong method, one unit of serum alanine aminotransferase (ALT) activity is defined as the amount of enzyme in one hundred milliliter serum catalyzing the formation of 1 micromole of pyruvic acid per sixty min under standard conditions.

The specific activity of an enzyme is another common unit, which is defined as the activity of an enzyme unit per milligram of total protein (expressed in $\mu mol\ min^{-1}\ mg^{-1}$). Specific activity gives a measurement of the purity of the enzyme.

## 4.2 Enzyme kinetics (Factors affecting reaction velocity)

The enzyme kinetics is aimed to study the rate of enzyme-catalyzed reaction and the factors that affect the velocity, including substrate concentration, enzyme concentration, pH, temperature, inhibitor, activator, and so on. When one of the factors is studied, the others are held constant. The study on enzyme kinetics is benefical to understand the mechanisms by

### 4.2.1 Effect of substrate concentration on reaction velocity

which enzymes work, and also design drug (special inhibitors) to overcome infections, correct metabolic disorders and treat tumors.

The initial velocity of reaction ($V_i$) varies hyperbolically with substrate concentration. At low substrate concentration, $V_i$ is almost directly proportional to substrate concentration. With the increase of substrate concentration, $V_i$ is no longer proportional to substrate concentration. If the substrate concentration is continuously increased and is high enough, the active sites of enzyme will be saturated with the substrate and the reaction velocity will reach a maximum, that is, $V_i$ is independent of substrate concentration.

The Michaelis-Menten equation illustrates.

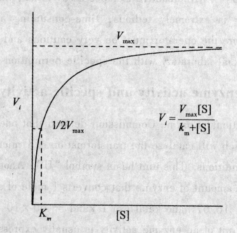

Fig. 4-1  the relationship between initial reaction velocity $V_i$ and substrate concentration [S] (Figwre 4-1)

The $V_{max}$ is the maximum velocity of the reaction. $K_m$ is the Michaelis constant. The $K_m$ of an enzyme is equal to the substrate concentration at which the reaction occurs at half of the maximum rate $V_{max}$. So, it is expressed in units of concentration, usually in molar units. $K_m$ is an indicator of the affinity that an enzyme has for a given substrate, and hence the stability of the enzyme-substrate complex. When the inhibitor of the enzyme existed, the type of inhibition can be determined by determining $V_{max}$ and $K_m$ for the enzyme.

There are limitations in the quantitative interpretation of a Michaelis plot. The $V_{max}$ is never really reached and therefore $V_{max}$ and hence $K_m$ values calculated from this graph are somewhat approximate. A more accurate way to determine $V_{max}$ and $K_m$ is to convert the data into a linear Lineweaver-Burk plot, which is a linear transformation of the Michaelis-Menton equation generated by taking the reciprocal of both sides of the equation. $V_{max}$ and $K_m$ are easily determined from the intercept on the Y or X axis.

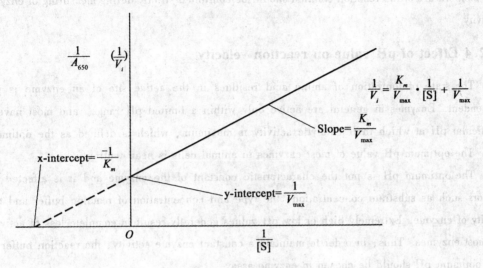

Fig. 4-2 Lineweaver-Burk Plot.

## 4.2.2 Effect of enzyme concentration on reaction velocity

In order to study the effect of the enzyme concentration on the reaction velocity, the substrate must be present in an excess amount, i. e., the reaction must be independent of the substrate concentration. Under this condition, the enzyme will be saturated with the substrate and the reaction velocity will be directly proportional to the enzyme concentration.

## 4.2.3 Effect of temperature on reaction velocity

Temperature displays dual influences on reaction velocity. The reaction velocity approximately doubles for every 10℃ rise in temperature. However, the higher temperature of reaction system could lead to a thermal denaturation of the enzyme and thus inactivate it. Taken together, the temperature at which the reaction velocity is the maximum is defined as the optimum temperature of an enzyme-catalyzed reaction, which is normally 35~40℃. However, the optimum temperature of an enzyme is not the characteristic constant of the enzyme and is time-dependent. Enzymes can tolerate the high temperature for a short time. Contrarily, the optimum temperature decreases as the reaction proceeds. For this reason, enzyme assays are routinely carried out at 30℃ or 37℃, but not always at the optimum temperature.

Although the enzyme activity decreases as the temperature declines, enzymes are generally not destroyed at low temperature. When temperature rises again, the enzyme activity recovers accordingly. To avoid the denaturation of enzyme, enzymes should be kept in refrigerator and be used immediately after taken out of refrigerator. In order to maintain the enzyme activity, the temperature of the reaction solution should be controlled strictly during measuring of enzyme activity.

### 4.2.4 Effect of pH value on reaction velocity

The state of ionization of amino acid residues in the active site of an enzyme is pH dependent. Enzymes in general are active only within a limited pH range, and most have a particular pH at which their catalytic activity is maximum, which is defined as the optimum pH. The optimum pH value of most enzymes in animal cells is near neutrality.

The optimum pH is not the characteristic constant of the enzyme and it is affected by factors such as substrate concentration, the type and concentration of reaction buffer and the purity of enzyme. Extremely high or low pH values generally result in complete loss of activity of most enzymes. Thus, in order to maintain a constant enzyme activity, the reaction buffer at the optimum pH should be chosen in enzyme assay.

### 4.2.5 Effects of inhibitor and activator on reaction velocity

Any substances that can convert an inactive enzyme into an active one or can increases the enzyme activity in a catalyzed reaction is called an enzyme activator. Enzyme activators can be classified as essential and non-essential activators. Most of the essential activators are metal ions, such as magnesium ($Mg^{2+}$), potassium ($K^+$) and manganese ($Mn^{2+}$), etc. They are essential for enzyme-catalyzed reactions and the enzyme activity can not be detected without these activators. Any substances that can reduce the enzyme activity without causing denaturation of enzyme are known as enzyme inhibitors. Based on either tightly or loosely bound to the enzymes, inhibitors can be divided into two types, the reversible inhibitors and the irreversible inhibitors.

## 4.3 Enzymatic analysis

Enzymatic analysis is an analytic method using enzymes as tools, based on which the quantity of substrates, coenzymes, activators or inhibitors can be measured. The commonly used methods include kinetic assay, end point assay and enzyme-linked immunosorbent assay (ELISA). The end point assay is the most common method in which sample is incubated with the buffered substrate for a fixed period of time and the amount of substrate used or product formed can be quantitatively measured when the reaction is stopped. End point assay can be divided into single enzymatic reaction assay and coupled enzymatic reaction assay.

Chapter IV Enzyme Analysis

## 4.3.1 Single enzymatic reaction assay

### 4.3.1.1 Measurement of the substrate amount

Single enzymatic reaction is defined as the reaction catalyzed by one enzyme. The reaction may be expressed as following formula:

$$S \xrightarrow{E} P$$

(1) Measurement of the substrate consumed

If the substance measured is a substrate in an enzyme-catalyzed reaction, and the substrate can be completely converted to product (when 99% of substrates are converted into products, the reaction is considered to be complete), then the substrate with a special character (such as a special absorption spectrum) can be directly determined by measuring the amount of substrate consumed in a reaction. For example, the amount of uric acid can be quantitatively measured by uricase. The reaction formula is shown as following:

$$\text{Uric acid} + 2H_2O + O_2 \xrightarrow{\text{Uricase}} \text{Allantion} + CO_2 + H_2O_2$$

Uric acid has two specific absorption peaks at wavelengths 293 nm and 297 nm, and their molar extinction coefficients are $12.6 \times 10^6$ and $11.7 \times 10^6$ respectively. Therefore, the amount of uric acid can be directly measured by monitoring the decrease of the absorbance at 293 nm and 297 nm in the reaction catalyzed by uricase.

(2) Measurement of the product formed

If the substance measured is a substrate and almost all substrates can be converted into products which can be specifically measured, then the amount of substrate can be measured by monitoring the amount of the product produced. For example, the quantification of hexose (microassay by using ATP-$\gamma$-[$^{32}$P]) is based on measurement of product.

$$\text{D-hexose} + \text{ATP-}\gamma - [^{32}P] \xrightarrow{\text{hexokinase}} \text{D-hexose-6-phosphate}[^{32}P] + \text{ADP}$$

In the reaction, the [$^{32}$P] of ATP-$\gamma$-[$^{32}$P] will be incorporated into D-hexose-6-phosphate through the catalysis of hexokinase. The amount of hexose can then be measured by monitoring the amount of [$^{32}$P] in this product.

(3) Determination of the amount of substrate by measuring the amount of coenzyme

In a reaction catalyzed by dehydrogenases utilizing $NAD^+$ or $NADP^+$ as coenzyme, the amount of substrate can be quantitatively measured by monitoring the absorbance change of NADH or NADPH at wavelength 340 nm (the molar extinction coefficient is $6.22 \times 10^6$). For example, the quantitative measurement of L-glutamic acid is based on measurement of the formation of coenzyme NADH.

$$\text{L-glutamic acid} \xrightarrow[\substack{\nearrow \\ NAD^+ + H_2O}]{\text{Glutamic acid dehydrogenase} \substack{\searrow \\ NADH + NH_4^+}} \alpha\text{-ketoglutarate}$$

### 4.3.1.2 Quantitative measurement of coenzymes

There are many kinds of coenzymes. Since coenzymes play important roles in maintaining the function of enzymes, it is necessary to quantitatively measure the coenzymes. Coenzyme can be measured by single enzymatic reaction. For example, CoA can be determined by using phosphate transacetylase (PTA). The reaction formula is shown below:

$$\text{CoA + acetyl phosphate} \xrightarrow{\text{Phosphate transacetylase}} \text{acetyl-CoA} + H_3PO_4$$

Since acetyl-CoA has an absorbance peak at wavelength 233 nm, the formation of acetyl-CoA can be determined by measuring the absorbance of acetyl-CoA at 233 nm and then the amount of CoA can be calculated based on the formation of acetyl-CoA.

### 4.3.2 Coupled enzymatic reaction assay

When the substrate or the product in single enzymatic reaction can't be measured by physical and chemical methods, the quantitative assay can be carried out by coupling another enzyme which acts as an indicator enzyme. For example,

$$A \xrightarrow{E_1} B \xrightarrow{E_2} C$$

In the enzyme-catalyzed reaction above, if A or B is difficult to be measured directly, then the amount of A can be determined by measuring the amount of C using $E_2$ as an indicator enzyme. The coupled reactions can be divided into two classes.

#### 4.3.2.1 The method using dehydrogenases as indicator enzymes.

The most commonly used indicator enzymes are dehydrogenases using $NAD^+$ or $NADP^+$ as coenzyme. For example,

$$\text{Glucose} \xrightarrow[\text{ATP} \quad \text{ADP}]{\text{Hexokinase}} \text{Glucose-6-phosphate}$$

$$\text{Glucose-6-phosphate} \xrightarrow[\text{NADP}^+ \quad \text{NADPH} + \text{H}^+]{\text{Glucose-6-phosphate dehydrogenase}} \text{6-phosphogluconate}$$

Because the specificity of hexokinase is not very high, the amount of glucose is difficult to be measured directly by single enzyme reaction. In coupled enzymatic reaction above, glucose-6-phosphate, the product of the first reaction, is used as substrate for the second reaction catalyzed by the indicator enzyme, glucose-6-phosphate dehydrogenase. The amount of NADPH produced in the coupled reaction is then easily determined by measuring the absorbance of NADPH at wavelength 340 nm and thus the amount of glucose can be quantitatively measured based on the NADPH produced.

#### 4.3.2.2 The method using other enzymes as indicator enzymes

Except for dehydrogenases, some enzymes can also be used as indicator enzymes in coupled enzymatic reaction. For example, glucose is oxidized by the enzyme glucose oxidase (GOD) to produce gluconic acid and release hydrogen peroxide ($H_2O_2$). $H_2O_2$ is further converted to water and oxidized products by coupling with the enzyme peroxidase (POD), which acts as an indicator enzyme.

$$\text{Glucose} + H_2O + O_2 \xrightarrow{\text{Glucose oxidase}} \text{Gluconic acid} + H_2O_2$$

$$\underset{\text{(reduced form of pigment)}}{H_2O_2 + DH_2} \xrightarrow{\text{Peroxidase}} \underset{\text{(oxidized form of pigment)}}{2H_2O + D}$$

The oxidized form of pigments such as dianisidine (DAD) and diaminobenzidine (DAB) have absorption peaks at wavelengths range from 420 nm to 470 nm. Therefore, the amount of glucose can be determined by measuring the formation of DAD/DAB by spectrophotometry.

## Experiment 9  Assay the Activity of Alanine Aminotransferase (ALT) in Serum (Mohun's Method)

### Principle

The catalytic activities in the plasma enzymes may serve as qualitative or quantitative indexes of tissue damage. Activities of serum alanine aminotransferase (ALT) were determined colorimetrically according to Mohun (1957).

ALT can catalyze the transamination between L-alanine and α-ketoglutarate, the equation is:

$$\text{Alanine} + \alpha\text{-ketoglutarate} \xrightarrow{\text{ALT}} \text{Pyruvate} + \text{Glutamic acid}$$

Then 2,4-dinitro-phenylhydrazine is added for preventing the reaction and marron compounds are formed which response to α-ketoacid. The absorbance of the product formed from pyruvate is bigger than that from α-ketoglutarate at 520 nm. So we can detect the activity of the ALT by measuring the absorbance of the colored product at 520 nm, which represent the amount of pyruvate produced in the reaction.

In this experiment, one unit of ALT activity is defined as the amount of enzyme needed to produce 2.5 μg of pyruvate per ml serum after it is incubated with the substrate at 37℃, pH 7.4 for 30 min.

### Reagents

1. Standard pyruvate solution (200 μg/ml, in 0.1 mol/L phosphate buffer, pH 7.4).
2. Substrate buffer (L-alanine and α-ketoglutarate).
3. 0.1 mol/L phosphate buffer (pH 7.4).
4. 0.4 mol/L NaOH.
5. 2,4-dinitro-phenylhydrazine solution (0.2 g/L in 1 mol/L HCl, dissolved by heating, stored in brown bottle, 4℃).
6. Sample: Serum, because hemolysis can cause a mild increase in ALT activity, sample collection and handling is important. Lipidemia also can cause an artificial increase in ALT activity in endpoint or nonkinetic assays. Improper sample handling leading to enzyme degradation may result in artificially decreased ALT values. Serum samples can be stored up

## Chapter IV  Enzyme Analysis

to 24 ~ 48 h in a refrigerator without loss of ALT activity. The ALT activity of serum samples is stable, as long as the serum is not frozen and thawed repeatedly.

## Procedure

1. 4 test tubes are used and the operation is done according to the following table.

| Reagents (ml) | Test (1) | Test Blank (2) | Standard (3) | Standard Blank (4) |
|---|---|---|---|---|
| Substrate buffer | 0.5 | - | 0.5 | 0.5 |
| Incubate the tubes in water bath at 37℃ for 5 min | | | | |
| Serum | 0.1 | 0.1 | - | - |
| Pyruvate (200 μg/ml) | - | - | 0.1 | - |
| Phosphate buffer | - | - | - | 0.1 |
| Mix the tubes, and incubate them in water bath at 37℃ for 30 min | | | | |
| 2,4-dinitro-phenylhydrazine | 0.5 | 0.5 | 0.5 | 0.5 |
| Substrate buffer | - | 0.5 | - | - |
| Mix the tubes, and incubate them in water bath at 37℃ for 20 min | | | | |
| 0.4mol/L NaOH | 5.0 | 5.0 | 5.0 | 5.0 |

Mix the tubes, after 20 min at room temperature, the absorbance is read within 30min at wavelength 520 nm on condition that absorbance is adjusted to zero with distilled water.

## Calculation

$$\text{ALT enzyme activity} = \frac{A_1 - A_2}{A_3 - A_4} \times \frac{20}{2.5} \times \frac{1}{0.1}$$

## Clinical significance

Plasma normally contains a number of specific enzyme molecules. These enzymes can be grouped as functional enzymes such as thrombin, lipoprotein lipase, etc., and nonfunctional enzymes, which include enzymes of exocrine secretions and intracellular enzymes, such as amylase, transaminase and lactate dehydrogenase. The plasma concentration of most enzymes remains constant in a normal individual. It will be altered if there is: a) change of synthesis of enzymes within the cell; b) cellular damage; c) change in the size of enzyme forming tissue; d) an alteration in the rate of inactivation and disposal of enzymes; e) an obstruction to a normal pathway of enzyme excretion.

Measurement of serum levels of numerous enzymes is of diagnostic significance. The serum nonfunctional enzyme determinations are particularly helpful in clinical medicine, it may be useful to: a) assess the severity of the organ damage; b) differentiate a particular type of disease; c) follow the trend of the disease; d) determine post operative risk.

The serum transaminase levels are normally low but elevated after extensive tissue destruction. Hepatic cell injury is manifested by elevated serum transaminase activity prior to the appearance of clinical symptoms. Comparable elevations of both AST and ALT are highly characteristic of acute viral, toxic, or non-ethanol drug-induced hepatitis. Hepatocytes contain more ALT than AST, so ALT is a more specific and sensitive indicator of liver damage.

Elevation of the serum transaminase levels may occur in a variety of non-hepatobiliary disorders. However, elevations exceeding 10 ~ 20 times the reference are uncommon in the absence of hepatic cell injury. Since the concentration of ALT is significantly less than AST in all cells except hepatocytes, ALT serum elevations are less common in non-hepatic disorders. Following myocardial infarction, AST activity is consistently increased. AST and only occasionally ALT increase in inflammatory skeletal muscle diseases and progressive muscular distrophy.

Chapter IV  Enzyme Analysis

**Experiment title**
**Date**
**Observations and results**

# Discussion

1. What is the clinical significance for assaying the activity of alanine aminitranserase in serum?

**Teacher's remarks**
**Signature**
**Date**

Experimental Manual in Medical Biochemistry

## Experiment 10  Lactate Dehydrogenase (LDH) Analysis

### Part I  Assay the Activity of Lactate Dehydrogenase (LDH) in Serum

### Principle

Activity of serum lactate dehydrogenase (LDH) was measured by a spectrophotometric assay. Lactate dehydrogenase (LDH) is an oxidoreductase that catalyzes the interconversion of lactate and pyruvate, the equation is:

$$\text{Lactate} + \text{NAD}^+ \xrightarrow{\text{LDH}} \text{Pyruvate} + \text{NADH} + \text{H}^+$$

Then 2, 4-dinitro-phenylhydrazine is added for preventing the reaction and marron compounds are formed which is response to pyruvate. The intensity of the color formed is directly proportional to the enzyme activity.

In this experiment, one unit of LDH activity is defined as the amount of enzyme needed to produce 1.0 μmol of pyruvate per 100 ml serum after it is incubated with the substrate at 37℃, pH 10.0 for 15 min.

### Reagents

1. Standard Pyruvate solution (1 μmol/ml, in 0.05 mol/L sulfuric acid).
2. Substrate buffer (lactate, pH 10.0).
3. $\text{NAD}^+$ solution.
4. 0.4 mol/L NaOH.
5. 2, 4-dinitro-phenylhydrazine solution (0.2 g/L in 1 mol/L HCl, dissolved by heating, stored in brown bottle, 4℃).
6. Sample: Fresh serum without hemolysis.

### Procedure

4 test tubes are used and the operation is done according to the following table.

| Reagents (ml) | Test (1) | Test Blank (2) | Standard (3) | Standard Blank (4) |
|---|---|---|---|---|
| Substrate buffer | 0.5 | 0.5 | 0.4 | 0.5 |
| Serum | 0.05 | 0.05 | 0 | 0 |
| Pyruvate (1 μmol/ml) | 0 | 0 | 0.1 | 0 |
| Incubate the tubes in water bath at 37℃ for 5 min | | | | |
| $NAD^+$ | 0.1 | 0 | 0 | 0 |
| Distilled water | 0 | 0.1 | 0.15 | 0.15 |
| Mix the tubes, and incubate them in water bath at 37℃ for 15 min | | | | |
| 2,4-dinitro-phenylhydrazine | 0.5 | 0.5 | 0.5 | 0.5 |
| Mix the tubes, and incubate them in water bath at 37℃ for 15 min | | | | |
| 0.4 mol/L NaOH | 5.0 | 5.0 | 5.0 | 5.0 |

Mix the tubes, after 5 min at room temperature, the absorbance is read within 15 min at 440 nm on condition that absorbance is adjusted to zero with distilled water.

**Calculation**

$$\text{LDH enzyme activity} = \frac{A_1 - A_2}{A_3 - A_4} \times 200$$

**Clinical significance**

When disease or injury affects tissues containing LDH, the cells release LDH into the blood, where it is identified in higher than normal levels. LDH is often measured to evaluate the presence of tissue or cell damage. Myocardial tissue is rich in LDH. Therefore measurement of the serum levels of LDH can be used to ascertain the potential for myocardial cell damage.

# Part II  Assay the LDH Isoenzyme by Agarose Electrophoresis

## Principle

Agarose is a linear polymer of alternating residues of D and L-galactose joined by α-(1-3) and β-(1-4) glycosidic linkages. Agarose gels have a greater range of separation for macrobiomolecules: proteins or larger DNA fragments. Agarose gels are often run in a horizontal configuration in an electric field of constant strength and direction.

Isoenzymes are enzymes that have different molecular forms but catalyze the same chemical reaction. Five major LDH isoenzymes are found in different vertebrate tissues. Each LDH molecule is composed of four polypeptide chains, which can be classified into two types: M (for skeletal muscle) and H (for heart muscle). In alkaline buffer solution, enzyme proteins are negatively charged. They will migrate toward anode in an electric field. Because of the difference of net charge number and molecular weight, their migration velocities are distinct. The LDH with more "H-subunit" that has more acidic amino acid residues will have greater migration velocity. The LDH with more "M-subunit" will have smaller migration velocity.

In the experiment, agarose gel is used as the supporting medium of electrophoresis to separate LDH isoenzyme in an electric field. After electrophoresis, the agarose gel was incubated in the dye solution. LDH catalyze the removal of hydrogen atoms from lactate. The product $NADH + H^+$ reacts with the artificial hydrogen transmitter phenazine methosulfate (PMS) and finally with the last hydrogen acceptor nitroblue tetrazolium (NBT). NBT is deoxidized to purple compound. Then every zones of LDH isoenzyme can be seen. Observe and judge the shade of color and width of the bands by naked eyes to estimate their quantity or estimate their percentages by spectrophotometer scan.

## Reagents

1. The electrophoresis buffer: 0.06 mol/L barbital sodium-barbital (pH 8.6). Take 12.76 g barbital sodium and 1.66 g barbital and dissolve them in 1L distilled water.
2. Dye solution: Prepared with water. The solution includes 211 mmol/L lactic acid sodium, 5.58 mmol/L coenzyme I, 0.4 mmol/L phenazine methosulfate (PMS) and 2.67 mmol/L nitroblue tetrazolium (NBT).
3. 0.5% agarose gel.

## Procedure

1. Prepare an agarose gel, take a flask containing 0.5% agarose gel and heat it in a microwave for 2~4 min until the agarose is dissolved. Pour the gel onto a slide, make a well with a plastic lid at the site 2 cm from cathode. Allow 20~30 min for solidification.
2. Carefully remove the plastic lid and load 20 μl sample into the wells. Place the gel in a horizontal electrophoresis apparatus. Connect the electrode and the gel with filter paper. Turn on the power and electrophoresis is carried out at a constant voltage of 90 V. The electrophoresis continues for 50~60 min.
3. After electrophoresis, take out the gel, place it in a big plate. Then add dye solution on the gel, keep it in darkness in a water incubator at 37℃ for 30 min or at 45℃ for 15 min.
4. Observe the shade of color and width of the bands by naked eyes to estimate their quantity.

## Reference

LDH isoenzyme in healthy adult serum has the order as follows:

$LDH_2 > LDH_1 > LDH_3 > LDH_4 > LDH_5$

The 5 types of LDH and their normal distribution levels are listed below.

$LDH_1$-Found in heart and red-blood cells and is 17% ~ 27% of the normal serum total.

$LDH_2$-Found in heart and red-blood cells and is 27% ~ 37% of the normal serum total.

$LDH_3$-Found in a variety of organs and is 18% ~ 25% of the normal serum total.

$LDH_4$-Found in a variety of organs and is 3% ~ 8% of the normal serum total.

$LDH_5$-Found in liver and skeletal muscle and is 0% ~ 5% of the normal serum total.

## Clinical significance

When LDH is released from damaged tissue, the LDH isoenzyme zymogram in the serum will change correspondingly as following:

1. $LDH_1$ and $LDH_2$ increase, $LDH_3$, $LDH_4$ and $LDH_5$ decrease
   (1) Tissue damage, such as acute myocardial infarction, myocarditis, etc.
   (2) Illness related to hemolysis, such as pernicious anemia, autoimmune hemolytic anemia, thalassemia, etc.
2. All five isoenzymes increase, especially $LDH_5$
   (1) Hepatic injury: hepatitis and cholestasis.
   (2) Injury of skeletal muscle.
   (3) Hypoxia.
3. $LDH_2$, $LDH_3$ and $LDH_4$ increase notably
   (1) Necrosis and inflammation of lung, pancreas, spleen, lymph node and the salivary gland.
   (2) Thrombocytosis or the tissue thrombogen kinase type substance enters the blood stream and causes the damage.
4. $LDH_1$ increases, and $LDH_5$ increases notably

   Appear in the diseases that heart and liver are involved in simultaneously such as congestive heart failure.

Experimental Manual in Medical Biochemistry

**Experiment title**
**Date**
**Observations and results**

**Discussion**

1. Describe briefly the clinical significance of the detection of LDH isoenzyme.
2. Explain the principle of the staining of LDH isoenzyme in the experiment.

**Teacher's remarks**
**Signature**
**Date**

# Experiment 11  Effect of Substrate Concentration on Enzyme Activity—Determining $K_m$ Value for Alkaline Phosphatase

## Principle

In this experiment we measure the value of the $K_m$ of alkaline phosphatase (ALP). Alkaline phosphatase catalyzes the following reaction:

$$\text{Disodium phenyl phosphate} + H_2O \xrightarrow[\text{pH10.0,37°C}]{\text{Alkaline phosphatase}} \text{Hydroxybenzene} + Na_2HPO_4$$

The concentration of hydroxybenzene can be obtained through the following reaction:

$$\begin{array}{c}\text{Hydroxybenzene} + \text{Phosphomolybdic acid-}\\ \text{phosphotungstic acid in phenol reagent}\end{array} \xrightarrow[OH^-]{\text{reduction}} \text{blue compounds}$$

The intensity (Absorbance, $A$) of the blue compounds correlates directly with the concentration of the substrate (disodium phenyl phosphate) and can be measured at 650nm. The difference of the concentration of disodium phenyl phosphate before and after the reaction equals to the reaction velocity ($V$). That is, $1/A_{650}$ equal to $1/V$. Plotting $1/A_{650}$ versus $1/[S]$ gives the Lineweaver-Burk plot whose x-intercept is $-1/K_m$, whose y-intercept is $1/V_{max}$, and whose slope is $K_m/V_{max}$.

## Reagents

1. Phenol reagent (prepared as the same in Lowry method).
2. 2.5 mmol/L disodium phenyl phosphate (substrate solution).
3. Alkaline buffer (pH 10.0):
   Sodium carbonate ($Na_2CO_3$) 6.36 g and sodium bicarbonate 3.36 g are dissolved in 1 L distilled water ($dH_2O$).
4. Alkaline phosphatase solution:
   Alkaline phosphatase 1 mg is dissolved in 4 ml $dH_2O$ (25 mg%).

Experimental Manual in Medical Biochemistry

## Procedure

Six test tubes are used and the operation is done according to the following table.

| Reagents (ml) | 1 | 2 | 3 | 4 | 5 | blank |
|---|---|---|---|---|---|---|
| 2.5 mmol/L Disodium phenyl phosphate | 0.2 | 0.4 | 0.6 | 0.8 | 1.0 | 1.0 |
| $dH_2O$ | 0.8 | 0.6 | 0.4 | 0.2 | - | - |
| Alkaline buffer | 1.0 | 1.0 | 1.0 | 1.0 | 1.0 | 1.0 |
| Mix the contents of tubes, incubate at 37℃ for 5 min | | | | | | |
| Alkaline phosphatase | 0.1 | 0.1 | 0.1 | 0.1 | 0.1 | - |
| Mix the contents of tubes, incubate at 37℃ for 15 min (Accurate!) | | | | | | |
| Phenol reagent | 1.0 | 1.0 | 1.0 | 1.0 | 1.0 | 1.0 |
| Alkaline phosphatase | - | - | - | - | - | 0.1 |
| 10% $Na_2CO_3$ | 3.0 | 3.0 | 3.0 | 3.0 | 3.0 | 3.0 |

Each tube must be mixed immediately after phenol reagent is added. After incubation at 37℃ for 15 min, the absorbance at wavelength 650 nm is detected on condition that blank tube is adjusted to zero.

## Calculation

1. To make the graph, take the reciprocal of each of your data points or in other words for each $A_{650}$ you calculate $1/A_{650}$ and for each [S] you calculate $1/[S]$.

| Tube | $A_{650}$ | $1/A_{650}$ | [S] | $1/[S]$ |
|---|---|---|---|---|
| 1 | | | | |
| 2 | | | | |
| 3 | | | | |
| 4 | | | | |
| 5 | | | | |

2. Prepare the Lineweaver-Burk plot by plotting $1/A_{650}$ versus $1/[S]$. $V_{max}$ and $K_m$ can be determined from the intercept on the Y or X axis.

Chapter IV  Enzyme Analysis

**Experiment title**
**Date**
**Observations and results**

**Discussion**

1. The $V_{max}$ is _____. The $K_m$ of an enzyme is the substrate concentration at which the reaction occurs at _____ of the maximum rate.

2. $K_m$ is (roughly) an inverse measure of the affinity or strength of binding between the enzyme and its substrate. The lower the $K_m$, the _____ the affinity.

**Teacher's remarks**
**Signature**
**Date**

85

Experimental Manual in Medical Biochemistry

## Experiment 12  Coupled Enzymatic Reaction Assay (Enzymatic Endpoint Method)-Determination of Total Cholesterol, Triglycerides and Glucose in Serum

### Part I  Determination of Total Cholesterol in Serum (CHO-PAP Enzymatic Endpoint Method)

**Principle**

The cholesterol is determined after enzymatic hydrolysis and oxidation. Cholesterol esters in the serum are hydrolyzed to free cholesterol and fatty acids in a reaction catalyzed by cholesterol esterase. Then the free cholesterol (the produced and already present in the serum) is oxidized by cholesterol oxidase forming hydrogen peroxide. The indicator quinone imine is formed from hydrogen peroxide and 4-aminoantipyrine in presence of phenol and peroxidase. The amount of total cholesterol (TC) in the sample is determined by measurement of the absorbance of the red color at 460~560 nm.

$$\text{Cholesterol ester} + H_2O \xrightarrow{\text{Cholesterol exterase}} \text{Cholesterol} + \text{Fatty acid}$$

$$\text{Cholesterol} + O_2 \xrightarrow{\text{Cholesterol oxidase}} \text{Cholestene-3-one} + H_2O_2$$

$$2H_2O_2 + \text{Phenol} + 4\text{-aminoantipyrine} \xrightarrow{\text{Peroxidase}} \text{Quinone imine} + H_2O$$

**Reagents**

1. Sample (Serum, heparinized plasma or EDTA plasma)
   Sample for lipid profile should be taken after 12 h fasting. This is because the chylomicron present in postprandial plasma elevates triglyceride and reduces LDL and HDL cholesterol. The chylomicrons are usually cleared after 6~9 h and their presence after 12 h is considered abnormal.
2. Standard cholesterol solution (200 mg/dL or 5.17 mmol/L).
3. Reaction reagent (pH 6.5, 30 mmol/L phosphate buffer containing Cholesterol esterase > 150 U/L, Cholesterol oxidase > 100 U/L, Peroxidase > 5000 U/L, 4-aminoantipyrene

0.25 mmol/L, Phenol 25 mmol/L) (Leadman kit).
4. Distilled water (dH$_2$O).

## Procedure

1. Accurately pipette each of the solutions into duplicate (or triplicate) wells in a microplate.

| Reagent (μl) | Blank | Test | Standard |
|---|---|---|---|
| Distilled H$_2$O | 5 | - | - |
| Standard solution | - | - | 5 |
| Sample | - | 5 | - |
| Reaction reagent | 200 | 200 | 200 |

2. Mix the wells by pipetting up and down several times. Incubate at 20℃ to 25℃ for 10 min or at 37℃ for 5 min (the developed color remains unchanged at least 1 h). Avoid exposure to direct light. Measure the absorbance at 500 nm of the test sample and the standard against the blank using a spectrophotometric microplate reader.

## Calculation

$$\text{Concentration of total cholesterol} = \frac{A_{\text{Test}}}{A_{\text{Standard}}} \times 200 \, (\text{mg/dL})$$

## Clinical significance

| Value | Interpretation |
|---|---|
| < 5.17 mmol/L (200 mg/dL) | Desirable blood cholesterol |
| 5.17 ~ 6.18 mmol/L (200 ~ 239 mg/dL) | Border-line high blood cholesterol |
| ≥ 6.20 mmol/L (240 mg/dL) | High blood cholesterol |

Cholesterol in the blood is present mostly in HDL (high density lipoprotein) and LDL (low density lipoprotein) in form of cholesterol ester and cholesterol.

### HDL cholesterol (HDL-C)

HDL cholesterol is measured in the supernatant after the precipitation of apoB-100 containing lipoprotein (LDL) directly from plasma or serum using a polyanion in presence of divalent cation.

### LDL cholesterol (LDL-C)

Total cholesterol(TC) = [VLDL cholesterol] + [LDL cholesterol] + [HDL cholesterol]
VLDL cholesterol(VLDL-C) = [plasma Triglyceride]/ 5 mg/dL
LDL cholesterol = [TC]-[HDL cholesterol]-[Triglyceride]/ 5 mg/dL

Experimental Manual in Medical Biochemistry

## 1. Lipid profile

Lipid profile is a measure of the lipid contents of the blood. It includes measurement of triglycerides, cholesterol (TC, HDL-C, LDL-C, VLDL-C). Cholesterol and triglycerides are insoluble in water and are transported in the blood in lipoproteins, complexes of lipids with specific proteins known as apolipoproteins. There are four major classes of lipoprotein: chylomicrons (CM), VLDL, LDL and HDL. It assumes major significance in diagnosis and management of cases of dyslipoproteinemia.

## 2. Dyslipoproteinemia

The causes of dyslipoproteinemia can be genetic or secondary due to a variety of other conditions, including nephrotic syndrome, obesity, alcoholism, hypothyroidism, diabetes mellitus and certain drugs. The clinical complication of such lipid and lipoprotein metabolism disorder can be:

(1) Premature atherosclerosis.
(2) Deposition of lipid into various tissues including dermis (xanthomas).
(3) Elevated triglycerides presents with a risk of pancreatitis.

Hyperlipoproteinemia may be classified into six distinct phenotypes (the WHO classification) according to lipoprotein particles present in excess.

| | | |
|---|---|---|
| I | Familial lipoprotein lipase deficiency | TG ↑↑ (due to chylomicron presence) |
| IIa | Familial hypercholesterolemia | LDL ↑ |
| IIb | Familial hypercholesterolemia | LDL ↑ and VLDL ↑ |
| III | Familial type III hyperlipoproteinemia | Chol ↑, TG ↑ (presence of β-VLDL) |
| | | VLDL Chol/ Plasma TG > 0.3 |
| IV | Familial hypertriacylglycerolemia | VLDL ↑ |
| V | Familial type V hyperlipoproteinemia | VLDL ↑ & presence of Chylomicron |

The most important primary hyperlipidemia is familial hypercholesterolemia. The molecular basic of this condition is a functional defection, or a decrease in the number of LDL receptors, which leads to a decrease in clearance of these lipoproteins from the blood, and an increase in cholesterol synthesis.

## 3. Screening for CHD (Coronary Heart Disease)

Cholesterol level is an important player in atherogenesis and consequently coronary heart disease (CHD). In particular LDL-C is implicated in atherogenesis and thus termed 'bad cholesterol'. HDL-C (involved in reverse cholesterol transport from the tissues to liver for excretion) is considered beneficial and rightly named 'good cholesterol'. The ratio of serum TC to HDL-C has a direct correlation with the risk factor of CHD. The ratios over 5:1 in male and 4.5:1 in female are considered as high risk of CHD. Therefore in screening for CHD it is necessary to measure HDL-C as well. Since serum lipids value varies from day to day, at least 2~3 measurements should be made days/ weeks apart. Experts agree on TC/ HDL-C ratios as a better assessment of individual risk of CHD.

## Part II  Determination of Triglycerides in Serum (GPO Enzymatic Endpoint Method)

### Principle

The triglycerides (TG) in the sample are hydrolyzed by lipase to release free fatty acids and glycerol. Glycerol is phosphorylated by ATP to glycerol-3-phosphate (G-3-P) in a reaction catalyzed by glycerol kinase (GK). G-3-P is oxidized to dihydroxyacetone phosphate catalyzed by the enzyme glycerol phosphate oxidase (GPO). In this reaction hydrogen peroxide ($H_2O_2$) is produced in equal molar concentration to the level of TG present in the sample. The indicator is a red dye, quinone imine, formed from hydrogen peroxide, 4-aminoantipyrine and 4-chlorophenol under the catalysis of peroxidase. The concentration of TG in the serum is determined by measurement of the absorbance at 460 ~ 560 nm.

$$\text{Triglyceride} + H_2O \xrightarrow{\text{Lipase}} \text{Glycerol} + \text{Fatty acid}$$

$$\text{Glycerol} + \text{ATP} \xrightarrow{\text{GK}} \text{Glycerol-3-phosphate} + \text{ADP}$$

$$\text{Glycerol-3-phosphate} + O_2 \xrightarrow{\text{GPO}} \text{Dihydroxyacetone phosphate} + H_2O_2$$

$$2H_2O_2 + \text{4-aminoantipyrine} + \text{P-chlorophenol} \xrightarrow{\text{Peroxidase}} \text{Red quinone} + 4H_2O$$

### Reagents

1. Sample (Serum, heparinized or EDTA plasma, obtained after an overnight fasting, the TG in the sample are stable for several days when stored in the refrigerator, storage of the sample at room temperature is not recommended, since phospholipids may hydrolyze and release free cholesterol, which interferes the determination of TG in serum.)
2. Standard TG solution (100 mg/dL or 1.14 mmol/L)
3. Reaction reagent (pH 7.0, 40 mmol/L phosphate buffer containing lipases >150 U/L, Glycerol Kinase >0.4 U/ml, GPO >1.5 U/ml, ATP 1 mmol/L, Peroxidase >5000 U/L, 4-aminoantipyrene 0.25 mmol/L, Chlorophenol 25 mmol/L) (Leadman kit)
4. Distilled water ($dH_2O$)

### Procedure

1. Accurately pipette each of the solutions into duplicate (or triplicate) wells in a microplate:

Experimental Manual in Medical Biochemistry

| Reagent (μl) | Blank | Test | Standard |
|---|---|---|---|
| Distilled H$_2$O | 5 | - | - |
| Standard solution | - | - | 5 |
| Sample | - | 5 | - |
| Reaction reagent | 200 | 200 | 200 |

2. Mix the wells by pipetting up and down several times. Incubate at 20℃ to 25℃ for 10 min or at 37℃ for 5 min. Avoid exposure to direct light. Measure the absorbance at 500 nm of the test sample and the standard against the reagent blank within 60 min using a spectrophotometric microtiter plate reader.

## Calculation

$$\text{Concentration of total triglyceride} = \frac{A_{Test}}{A_{Standard}} \times 100 \, (\text{mg/dL})$$

## Clinical significance

### Reference Values

The following upper limits are recommended for the determination of the risk factor of hypertriglyceridemia:

Suspected form      --1.71 mmol/L or 150 mg/dL

Increased form      --2.29 mmol/L or 200 mg/dL

Triglycerides (TG) are transported in the blood as core constituents of all lipoproteins, but the greatest concentration of these molecules is carried in the TG-rich chylomicrons (CM) and very low density lipoproteins (VLDL). CM ultimately enter the blood compartment, where most of the TG is rapidly hydrolyzed in the capillary bed by lipoprotein lipase into glycerol and free fatty acids, which are absorbed by adipose tissue for storage or by other tissues requiring a source of energy. A peak concentration of CM-associated TG occurs within 3 ~ 6 h after ingestion of a fat-rich meal. CM that have been hydrolyzed in the circulation are termed CM remnants. They are taken up by the liver through a receptor-mediated process in which apoE and/or apoB on the CM remnant surface binds to the apoE receptor, LDL receptor or LDL receptor-related protein (LRP). Significant amounts of circulating TG are also transported in VLDL. After synthesis in the liver and secretion into the blood stream, they are hydrolyzed by LRP into VLDL remnants. Then, the majority continues to be hydrolyzed to LDL.

TG measurements are used in the diagnosis and treatment of diseases involving lipid metabolism (hyperlipoproteinemia), various endocrine disorders (diabetes mellitus, nephrosis) and liver obstruction (extrahepatic biliary obstruction). Although the relationship between TG concentration and risk of CHD has not been firmly established, several studies have found increased TG concentrations to be positively correlated with increased risk for CHD. However, whether TG concentration represents an independent risk factor is not clear.

TG is determined by levels of plasma lipids after an overnight fasting. Elevations of plasma TG may be due to increased levels of VLDL or combination of VLDL and chylomicron. Only when plasma TG concentration >11 mmol/L plus eruptive xanthomas, papules appearance and pancreatitis, high TG represents a major risk. Hypertriglyceridemia is commonly associated with obesity, excessive consumption of simple sugars and saturated fats, inactivity, alcohol consumption, insulin resistance and renal failure.

## Part III  Determination of Blood Glucose (Glucose Oxidase-Peroxidase Endpoint Method / GOD-PAP)

### Principle

Glucose is oxidized by glucose oxidase forming gluconic acid and hydrogen peroxide. The hydrogen peroxide formed is broken down by peroxidase to water and oxygen. The latter oxidizes phenol, which combines with 4-aminoantipyrine to give quinone imine, a red colored complex. The intensity of the red colored complex measured at 500 nm (460 ~ 560 nm) is proportional to the concentration of glucose in the sample.

$$\text{Glucose} + H_2O + O_2 \xrightarrow{\text{Glucose oxidase}} \text{Gluconic acid} + H_2O_2$$

$$H_2O_2 + \text{4-aminoantipyrine} + \text{phenol} \xrightarrow{\text{Peroxidase}} \text{Quinone imine} + H_2O$$

### Reagents

1. Sample (serum, or plasma anti-coagulated by heparin or EDTA).
   The serum or plasma should be separated from the whole blood within 1 h after collection and stored at 2 ~ 8 ℃ less than 24 h.
2. Working reagent (glucose oxidase > 15 U/ml, peroxidase > 1.5 U/ml, 4-aminoantipyrine 0.25 mmol/L, phenol 0.75 mmol/L).
   The dispensed compound may store 4 ~ 8 ℃ for one month. (Leadman kit).
3. Glucose standard solution (5.55 mmol/L, or 100 mg/dL).
4. Distilled water ($dH_2O$).

## Procedure

1. Accurately pipette each of the solutions into duplicate (or triplicate) wells in a microplate.

| Reagent (μl) | Blank | Test | Standard |
|---|---|---|---|
| dH$_2$O | 5 | - | - |
| Standard solution | - | - | 5 |
| Sample | - | 5 | - |
| Working reagent | 200 | 200 | 200 |

2. Mix the wells by pipetting up and down several times. Incubate them in a water bath at 37℃ for 15 min. Avoid exposure to direct light. Measure the absorbance of test and standard at 500 nm within 30 min using a spectrophotometric microplate reader.

## Calculation

$$\text{Concentration of blood glucose} = \frac{A_{Test}}{A_{Standard}} \times 100 \, (\text{mg/dL})$$

If the concentration of blood Glucose > 27 mmol/L, please dilute the serum or plasma with the same volume of 0.9% NaCl at first and multiply the result of assaying with times of dilution.

## Reference Values

Concentration of glucose in fasting serum/ plasma: 70 ~ 110 mg/dL or 3.89 ~ 6.11 mmol/L.

## Clinical significance

Glucose is a major sugar in blood. Blood glucose level is strictly regulated by hormones. Insulin decreases blood glucose level, whereas other hormones such as epinephrine, growth hormone and glucocorticoid increase glucose level.

Blood glucose levels fluctuate physiologically as well as pathologically. The condition that blood level above the normal reference level is called hyperglycemia and that below the normal level is called hypoglycemia.

Determining the blood glucose level plays very important roles in identification of various glucose metabolism disorders such as diabetes mellitus.

Please read "Experiment 2" for more details.

Chapter IV  Enzyme Analysis

**Experiment title**
**Date**
**Observations and results**

**Discussion**

1. The main lipids in the blood are _____ and _____. They are transported in the blood as lipoproteins, which are complexes of lipids and _____.
2. Considering the relationship between lipoproteins and atherosclerosis, _____ is implicated in atherogenesis and thus termed 'bad lipoprotein'. _____ is involved in reverse cholesterol transport and rightly named 'good lipoprotein'.
3. What are the main lipoproteins in plasma?
4. What kind of lipoprotein has the highest triglyceride content?
5. Why is plasma TG level increased in diabetes mellitus?
6. Describe briefly the clinical significance of determining blood glucose level.

**Teacher's remarks**
**Signature**
**Date**

93

# Chapter V

# Isolation and Purification of Protein

Isolation and purification of various types of proteins and peptides is used in almost all branches of biosciences and biotechnologies. Purification of protein is a series of processes intended to isolate one particular protein of interest from some complex biological (cellular) material, and it may be analytical or preparative.

The purposes for which protein separations are carried out differ widely, and accordingly the criteria of purity also differ. For example, the investigation of the structure, composition or amino acid sequence of a protein requires a purified homogeneous sample. The study of protein's structure and function requires their native structures and biological activities are preserved. The preparation of a protein for commercial or biopharmaceutical use may require large quantities of stable, well-characterized products, at a low cost. Most protein separations require that several criteria be met, such as degree of purity, yield and economy of the method. The key to efficient protein purification is to select the most appropriate techniques, optimize their performance and combine them in a logical way to maximize yield and minimize the number of steps required.

This chapter discusses protein separation and characterization techniques and presents a guide to planning and evaluating a protein purification scheme.

## 5.1 Methods and Basic Principles of Protein Purification

Based on the protein's physicochemical properties, such as its size, shape, isoelectric point (pI), solubility and hydrophobicity, and some of its chemistry, one kind of protein can be isolated and purified from a complex mixture using a group of reasonable separation methods (Table 5-1). Usually more than one purification step is necessary to achieve the desired purity.

Chapter V Isolation and Purification of Protein

Table 5-1    **Methods for protein separation**

| Physicochemical Properties | Methods for Separation |
| --- | --- |
| Solubility | Precipitation by salting out (e.g. ammonium sulfate); Acetone; pI; PEG |
| Size and shape | Gel-electrophoresis; Gel filtration; Centrifugation; Dialysis and Ultrafiltration |
| Charge | Gel-electrophoresis; Ion exchange chromatography |
| Isoelectric point (pI) | Isoelectric focusing electrophoresis; chromatofocusing |
| Hydrophobicity | Hydrophobic interaction chromatography (HIC); Reverse phase-HPLC |
| Biospecific interaction | Affinity chromatography |

## 5.1.1 Separation by precipitation

Precipitation with salts such as ammonium sulfate at high ionic strengths or with organic solvents (e.g. ethanol) at pI is a very common initial step in protein purification. This operation also serves to concentrate and fractionate the target product from various contaminants. It has the advantages of economy, simplicity and high capacity.

### 5.1.1.1 Salting out

The solubility of protein depends on the salt concentration in the solution. Salt at low concentrations stabilizes the various charged groups on a protein molecule, and enhances the solubility of protein. This is commonly known as salting-in. However, when the salt concentration is increased further, it implies that there is less and less water available to solubilize protein. Finally, protein starts to precipitate when there are not sufficient water molecules. This phenomenon of protein precipitation at high salt concentration is known as salting-out.

Salting out is largely dependent on the hydrophobicity of the protein. A typical protein molecule in solution has hydrophobic patches on its surface, while hydrophilic amino acid interact with the molecules of solvation and allow proteins to form hydrogen bonds with the surrounding water molecules. When the salt concentration (ionic strength) is increased, some of the water molecules are attracted by the salt ions, then the protein-protein interactions are stronger than the solvent-solute interactions, the protein molecules aggregate by forming hydrophobic interactions with each other.

Many types of salts have been employed in salting out. Ammonium sulfate is commonly used as its high solubility and relatively inexpensive. The salt concentration needed for the protein to precipitate out of the solution differs from protein to protein.

### 5.1.1.2 Isoelectric precipitation

The isoelectric point (pI) of protein is the pH of a solution at which protein has no net charge. Most proteins have a minimum solubility around their pI because an overall charge near zero minimizes repulsive electrostatic forces and cause hydrophobic forces to attract molecules to each other. This is called isoelectric precipitation. The greatest disadvantage to isoelectric precipitation is the irreversible denaturation because it normally occurs at lower pH below physiological values ( the pI of most proteins is in the pH range of 4~6), thus this precipitation is most often used to precipitate contaminant proteins, rather than the target protein.

#### 5.1.1.3 Precipitation with organic solvents

Addition of a miscible solvent such as ethanol, methanol or acetone to a solution may cause proteins in the solution to precipitate. The principle effect is the reduction of the dielectric constant of water as the concentration of organic solvent increases. With smaller hydration layers, the proteins can aggregate by attractive electrostatic and dipolar forces. Another feature affecting organic solvent precipitation is the size of the molecule. The larger the molecule, the lower percentage of organic solvent required to precipitate.

To avoid denaturation of protein in the less polar solvent, the precipitation is usually carried out at subzero temperatures. In operating procedure, the protein solution should be chilled close to 0℃ and the solvent to at least-20℃.

#### 5.1.1.4 Precipitation with organic polymers

Polyethylene glycol (PEG) is a hydrophilic polymer and will precipitate proteins without denaturing them. The behavior of the proteins is rather similar to their behavior in precipitation by organic solvents. PEG of molecular weights 6,000 and 20,000 is most often used for protein isolation. Large proteins are precipitated at lower PEG concentrations than small proteins.

After precipitation, PEG is not easy to remove from a protein fraction. As a polymer it will not dialyze rapidly. Acetone precipitation of the protein can be used to separate it from the PEG because PEG is acetone soluble. Another practical difficulty in the use of PEG is that it absorbs UV light at 280 nm, and also interferes with the Lowry protein assay. However, protein in the presence of PEG can be measured using the biuret assay.

#### 5.1.1.5 Affinity precipitation

Affinity methods in protein purification are defined as procedures that make use of the protein's specific and selective interaction with a ligand. These methods include affinity chromatography, affinity electrophoresis, affinity phase partitioning, and affinity precipitation, in which addition of a compound that specifically binds to the enzyme or protein, causes the target protein to aggregate and precipitate from solution.

Immunoprecipitation is a form of affinity precipitation. Addition of antibody (a specific ligand) to a protein solution, a giant molecular interaction aggregate is build up, and precipitation of protein can occur.

## 5.1.2 Dialysis and Ultrafiltration

Dialysis is the process of separating molecules in solution by the difference in their rates of diffusion through a semi-permeable membrane. Typically a semipermeable dialysis bag such as a cellulose membrane with pores, is loaded with a solution of several types of molecules and placed in a container of a dialysis solution. Small molecules and water tend to move into or out of the dialysis bag in the direction of decreasing concentration, while proteins have dimensions significantly greater than the pore diameter are retained inside the dialysis bag (Fig. 5-1A).

Ultrafiltration is a technique related to dialysis, using high-pressure by gas pressure or centrifugation, to force water and other small molecules through the selective permeable membrane, the proteins are retained and concentrated in the container. Ultrafiltration is mainly used as a rapid method for concentrating proteins, without expecting significant purification. Membranes with a nominal cutoff size of 3 kU to 100 kU are most frequently used, a proportion of molecules of sizes close to the cutoff size will pass through, the remainder staying inside. UF can be preformed using centrifuge concentrators or stirred-cell apparatus (Fig. 5-1B)

Dialysis and ultrafiltration are typically used to separate proteins from buffer components for buffer exchange, desalting, or concentration.

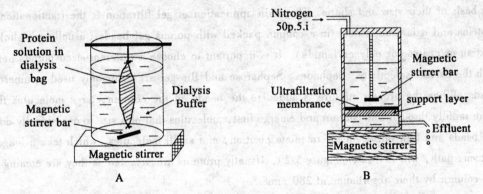

Fig. 5-1 Dialysis (A) and Ultrafiltration (B) of protein solution.

## 5.1.3 Ultracentrifugation

Ultracentrifugation is valuable for separating biomolecules and investigating such parameters as mass and density by calculating their sedimentation coefficient(S), and learning about the shape of a molecule. The most commonly density gradient ultracentrifugation is used to separate proteins with different S, especially useful for separation of lipoproteins.

Experimental Manual in Medical Biochemistry

## 5.1.4 Purification by Chromatography

Most purification schemes involve high resolution chromatographic techniques such as ion exchange chromatography, gel filtration, affinity chromatography, et al. (see chapter 3)

### 5.1.4.1 Ion-exchange chromatography

Ion-exchange chromatography (IEC) is one of the most popular selective purification methods. It separates proteins based on the reversible ionic interaction between a charged protein and an oppositely charged ion exchanger which is either polyanions or polycations. It is a very high resolution separation with high sample loading capacity. Proteins bind to charged matrix as they are loaded onto a column, then the elution is usually performed by stepwise or continuous gradient increases in salt concentration (ionic strength) or adjusts the pH of the solvent. Target proteins are collected in a purified, concentrated form.

Most proteins in buffer at physiologic pH carry a net negative charge and will bind to an ion exchanger that carries positively charged diethylaminoethyl (DEAE) groups. DEAE-Sephacel, DEAE-Sepharose CL-6B are most common ion exchangers, which have high capacity for proteins, excellent resolution and flow rate, and regeneration can be performed.

### 5.1.4.2 Size exclusion chromatography/ Gel filtration

Size exclusion chromatography (SEC), also called gel filtration, separates molecules on the basis of their size and shape. The main application of gel filtration is the fractionation of proteins and it is usually done in a column packed with porous gel beads (usually crosslinked dextran or agarose or polyacrylamide). It is important to choose a different-sized porous beads with the correct selectivity. Sephadex, Sepharose and Bio-gel are commonly used commercial beads. When the protein mixture is added to the beads on the column, large molecules flow more rapidly through this column and emerge first, molecules that of a size to occasionally enter the beads and flow out at an intermediate position, and small molecules, which take a longer, tortuous path, will exit last (Figure 5-2). Usually proteins are detected as they are coming off the column by their absorbance at 280 nm.

### 5.1.4.3 Affinity chromatography

Affinity chromatography (AC) is probably the most powerful of the chromatographic techniques. The separation of protein is based on a reversible interaction between a protein and a specific ligand attached to a chromatographic matrix (such as that between antigen and antibody, enzyme and substrate, or receptor and ligand). After loading the sample onto an affinity column, proteins specifically bind to a complementary binding substance. Unbound material is washed away, then desorption of the target protein is performed specifically, using a competitive ligand, or non-specifically, by changing the pH, ionic strength or polarity. Protein is collected in a purified, concentrated form.

Because affinity is dependent on biological activity, an affinity step will often discriminate

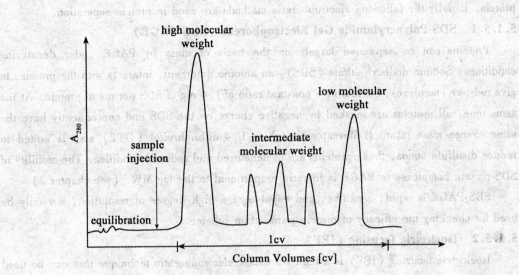

Fig5-2 Typical gel filtration elution.

between active and inactive forms of enzymes.

### 5.1.4.4 Hydrophobic interaction chromatography

Hydrophobic interaction chromatography (HIC) separates proteins with differences in hydrophobicity. The separation is based on the reversible interaction between a protein and the hydrophobic surface of a chromatographic matrix. A certain concentration of ammonium sulfate just below used for salting out is the most likely to promote binding of proteins to a hydrophobic gel. Fractionation of bound proteins is frequently achieved by decreasing the ionic strength or decreasing the polarity of the eluent (e.g. with ethylene glycol). Phenyl Sepharose CL-4B is an electrically neutral gel for HIC with hydrophobic groups attached by stable ether linkages.

### 5.1.4.5 High performance liquid chromatography

Using of a technique called high performance liquid chromatography (HPLC) can improve substantially the resolving power of all of the column chromatography with the use of conventional gravity flow column. In HPLC, the media is made up of finer material (e.g. non-compressible resins) and more strong columns is produced with metal, hence, pressures of 5000~10,000 psi are applied to force solutions rapidly through the column to obtain adequate flow rates. Because of high resolution of as well as rapid separation of HPLC, it is the most powerful and versatile form of chromatography.

## 5.1.5 Separation by Electrophoresis

Proteins are ampholytes having a pH-dependent net charge in a solution. Using electrophoresis in different ways, it is possible to separate proteins, detect protein contamination, estimate MW and pI and obtain information on the physiological activity of the protein. Usually the following electrophoretic methods are used in protein separation.

#### 5.1.5.1 SDS-Polyacrylamide Gel Electrophoresis (SDS-PAGE)

Proteins can be separated largely on the basis of mass by PAGE under denaturing conditions. Sodium dodecyl sulfate (SDS), an anionic detergent, interacts with the proteins to give rod-like complexes, containing a constant ratio of 1.4 mg of SDS per mg of protein. At the same time, all proteins are masked by negative charge on the SDS and consequently have the same charge/mass ratio. β-Mercaptoethanol or 1,4-dithiothreitol (DTT) also is added to reduce disulfide bonds, thus, proteins are in denatured and reduced condition. The mobility of SDS-protein complexes in PAGE is linearly proportional to the log MW. (see chapter 2)

SDS-PAGE is rapid, sensitive, and capable of a high degree of resolution, normally be used for checking the efficacy of protein purification scheme.

#### 5.1.5.2 Isoelectric focusing (IEF)

Isoelectric focusing (IEF) is a high resolution electrophoretic technique that can be used for preparative work as well as for analytical studies. Proteins are separated according to their charges (isoelectric points, pI). IEF can readily resolve proteins that differ in pI by as little as 0.01, which means that proteins differing by one net charge can be separated. Each protein will move until it reaches a position in the gel at which the pH is equal to the pI of the protein.

#### 5.1.5.3 Two-dimensional electrophoresis (2-DE)

Two-dimensional electrophoresis (2-DE), in which proteins are separated according to charge (pI) by IEF in the first dimension and according to size by SDS-PAGE in the second dimension, has a unique capacity for the very high resolution of protein mixtures. (see chapter 2) Proteins isolated from cells can be displayed to hundreds or even thousands of protein spots in a gel by 2-DE, followed by an examination of the intensity of the signals, then proteins identification by coupling with mass spectrometric techniques, such as matrix-assisted laser desorption-ionization (MALDI) or tandem mass spectrometry.

2-DE is the core technology of proteome analysis, as a good method of choice for protein micropreparative separation and spot identification.

## 5.2 Procedure of Protein Preparation

### 5.2.1 The Initial Preparation Step

It is possible to give detailed consideration to the initial preparation steps. Factors to consider at this stage are the choice of source material, extraction and crude fractionation. It is worthwhile to note some general precautions from beginning to the last step in a separation. For example, keeping temperature as low as possible or adding general antibiotics and proteinase inhibitors for retard the growth of micro-organisms and inhibit the activity of proteolytic enzymes; extremes of ionic strength and pH of solution, adding organic solvents and detergents or violent agitation should be avoided because they tend to denature proteins.

### 5.2.1.1 Choice of materials and pretreatment

The source from which the protein is isolated is the key to the design of a purification process. Choice of a starting raw material (such as bacteria, animal tissue/cell, and serum/plasma) is mainly determined by the experiment goal. Bacteria as material are normally used to overexpress a specific protein, polypeptide or enzyme for isolation. If selecting an animal active ingredient rich in tissues or cell, the crude materials should be fresh or frozen immediately before treatment, avoiding proteolytic degradation.

### 5.2.1.2 Cell disruption and protein extraction

Except the starting material is serum/plasma or culture supernatant, it is necessary to isolate the tissue/cells or sub-cellular fraction and disrupt the cellular material in buffer. Several methods such as repeated freezing-thawing, sonication and homogenization, are useful for disruption of cells (Table 5-2). The method of choice depends on how fragile the protein is and how sturdy the cells are. No matter what kind of methods used broken cells, the proteinase of cells would be released into the solution, causing proteins degradation. Therefore, some conditions are often required at this stage, such as adding diisopropyl fluoride phosphate (DFP), iodoacetic acid, phenylmethylsulfonyl fluoride (PMSF), or selecting a suitable pH, ionic strength or temperature.

Tab. 5-2    **Methods and their principles of disruption of cells**

| Methods | Principles |
| --- | --- |
| Low osmotic pressure | Osmotic disruption of cell membrane |
| Enzyme digestion (e.g. lysozyme) | Cell wall digested, leading to disruption of cell membrane |
| Chemical solubilization (e.g. SDS) | Cell membrane solubilized chemically |
| Homogenization | Cell forced through narrow gap, disrupt cell membrane but not organelles |
| Mincing/Grinding | Cell disrupted in grinding process by shear force |
| Bead mill | Rapid vibration with glass beads rip cell walls off |
| French press | Cell forced through small hole at high pressure, shear forces disrupt cells |
| Ultrasonication | Microscale high pressure sound waves causes cell breakage by shear forces |
| Repeated freezing-thawing | Cells ice formation and swelling, broken cell structure |

The extracts containing the soluble and particulate fractions are then clarified by filtration or centrifugation steps and checked for the distribution of protein. If the desired protein is associated with the particulate fraction, it will be necessary to solubilize it. The proteins in a soluble form are suitable for the next manipulation.

Most extracting proteins dissolve in salt solution and dilute acid or alkali buffer system, but some proteins such as lipoproteins dissolve in organic solvents such as ethanol, n-butanol and acetone. Organic solvent extraction is also the common method for purification of apolipoproteins.

### 5.2.1.3 Crude fractionation

The crude preparation after extraction is normally rather dilute, contains a large number of different proteins and has a large volume. The next step chosen normally is concentrated the sample for further processing.

Precipitation with organic solvents at pI or salting out is very common initial step in protein purification. It is of economy, simplicity and high capacity. However, it is not suitable for low molecular weight components which are very soluble. A useful alternative to precipitation is ion-exchange chromatography. If the charge of the target protein is known more, using an anion exchanger for doing the batch adsorption will be a very selective method.

Bioselective adsorption can also be used at this stage if something is known of the biological properties of the target protein. Although affinity gels are usually more expensive, the very high selectivity of affinity techniques, coupled with the high capacity of affinity chromatography media, makes it a powerful tool.

## 5.2.2 The Selective Purification Step

At this point the material of interest has been extracted from its source and the major of contaminants has been removed. Unless an affinity step has been used, the methods used so far have had high capacity but relatively poor selectivity. Since the final aim is to obtain the protein of interest in a suitably pure and active form for further study, the next steps need to be more selective. The choice of selective technique mainly depends on a consideration of the principle and compatibility of the techniques. It is not desirable to use any method which fractionates on the basis of a property more than once in any purification scheme. For example, IEC should not be followed by another technique according to the charge.

Gel filtration, hydrophobic interaction chromatography (HIC), ion-exchange chromatography (IEC) and affinity chromatography (AC) can be explored at this stage. Generally, gel filtration is a particularly suitable technique to reserve for late in the isolation when only a few components remain, since it is not a highly discrimination. For fractionation purposes it is important to choose a gel with the correct selectivity and the elution conditions which preserve the activity of the protein of interest. HIC is particularly applicable to samples which have a high salt concentration that just promote the binding of proteins to a hydrophobic gel. IEC is the one of the most popular selective methods and is particular suitable after gel filtration, that product has a rather large volume and may be dilute. It is important that the preceding step should be performed with a suitable ion exchanger and buffer. AC offers high selectivity and high capacity if the starting conditions are optimal for binding, and often achieves the desirable aims of a rapid isolation using a minimum number of steps, so it is an ideal method and always preferentially used whenever a suitable ligand is available for the protein of interest.

## 5.3 Crystallization, Concentration, Drying and Storage of Protein

### 5.3.1 Crystallization

Crystallization, the process of solidifying from solution, is often the final stage in purifying and studying proteins, particularly enzymes. If a solution saturated at some temperature is cooled, the dissolved component begins to separate from the solution. Because the solubilities of two solid compounds in a particular solvent generally differ, it often is possible to find conditions that only one of the components of a mixture crystallizes alone, while the more soluble components remain dissolved.

In order to obtain crystallized protein with a high purity, the crystallization should take place slowly. If solidification is rapid, impurities can be entrapped in the solid matrix. Sometimes, it is necessary to add a seed crystal to the solution to provide a solid surface for beginning the crystallization process.

Except as a method of purification, crystallization is also useful for confirmation of protein homogeneity, determining for tertiary structure of protein by X-ray diffraction techniques and storage of purified protein.

### 5.3.2 Concentration

Protein concentration is a useful step in purifying proteins. Usually it is necessary to concentrate a sample following gel filtration, and can improve the sensitivity of subsequent analysis. In addition, proteins are generally more stable in relatively concentrated solutions. Some methods can be employed as a means of concentration, and the optimal concentration method depends on the volume of the sample, the type of protein, the type of buffer employed and the aim of preparation.

#### 5.3.2.1 Concentrative dialysis

Absorbers can be used to remove solvent molecules so that samples can be condensed. The solution is placed into dialysis bag, surrounded by absorbers, e.g. sucrose, PEG or Bio-Gel concentrator resin.

#### 5.3.2.2 Ultrafiltration

Ultrafiltration by ultracent and minicent ultrafiltration cartridges provides an easy method for protein concentration.

#### 5.3.2.3 Vacuum evaporation/distillation

Lowering the vapor pressure at the given temperature lowers the boiling points of liquids, including that of the solvent. This allows the solvent to be removed without excessive heating. This technique is also called vacuum distillation and it is commonly carried out in the form of the rotary evaporator.

### 5.3.3 Drying

Drying protein is a dehydration process typically used for long term storage or more convenient for transportation. Normally, the methods used for protein drying have vacuum drying and freeze drying (also known as lyophilization). Vacuum drying is the final result of vacuum evaporation. Freeze drying can also be used as a late-stage purification step, because it can effectively remove solvents. Furthermore, it is capable of concentrating substances with low molecular weights that are too small to be removed by a filtration membrane.

### 5.3.4 Storage

Purified proteins often need to be stored for an extended period of time while retaining their original structural integrity and/or activity. Normally recommendations for storage conditions of purified proteins include (1) store in high concentration of ammonium sulfate (e.g. 4 mol/L); (2) freeze in 25% ~50% glycerol, especially suitable for enzymes; (3) sterile filter to avoid bacterial growth; (4) lyophilization. The extent of storage 'shelf life' can vary from a few days to more than a year and is dependent on the protein nature and the storage conditions. For example, proteins stored in solution at 4℃ can be dispensed conveniently but require more diligence to prevent microbial or proteolytic degradation; such proteins may not be stable for more than a few days or weeks. By contrast, lyophilization allows for long-term storage of protein (several years), but the protein must be reconstituted before use and may be damaged by the lyophilization process.

## 5.4 Identification and Analysis of Protein

For determining the effect of a protein purification scheme, one way is to measure the purity of protein, another is to evaluate yield and specific activity of protein at each purification step. In addition, the assessment of purity should be considered within the overall economy of the purification scheme, therefore, the methods should be quick and efficient.

### 5.4.1 Purity of Protein

Throughout the purification scheme it is important to check the purity of the samples. The most general method to monitor the purification process is by running a SDS-PAGE in the different steps. The components of the numbered peaks are examined, and the MW of proteins can also be determined by running a protein standards and plotting a calibration curve. However, this method only gives a rough measure of the amounts of different proteins in the mixture, and it is not able to distinguish between proteins with similar MW.

If the protein has a distinguishing spectroscopic feature, it can be used to detect and quantify the specific protein. If antibodies against the protein are available, then Western blotting and ELISA can specifically detect and quantify the amount of desired protein. Some proteins function as receptors and can be detected during purification steps by a ligand binding assay.

### 5.4.2 Purification Yield

In order to evaluate the process of multistep purification, the amount of the specific proteins has to be compared to the amount of total protein. Some of the available methods for measuring protein concentration are described previously, such as Biuret reaction, Lowry assay, Bradford assay, BCA method or UV absorption ($A_{280}$) (see chapter I experiment 1). However some reagents used during the purification process may interfere with the quantification. For example, imidazole (commonly used for purification of His-tagged recombinant proteins) is an amino acid analogue and will interfere with the BCA assay at low concentration.

### 5.4.3 Activity of Purified Protein

During a protein purification procedure, the most important thing to follow is the recovery of the protein by enzyme activity or bioactivity. If the protein is an enzyme, it required a method for determining the activity of enzyme. The use of zymogram techniques enables detailed information of enzyme activity. For detecting the immunological activity of the protein, immunoelectrophorosis is a good technique that is extremely useful for quantitative measurement, and also for immunological status.

### 5.4.4 Analysis of Purified Protein

Purification of protein is the basis of further study of structure and function, or as commercial or biopharmaceutical use. As mentioned above, there are some specific methods that directly apply to large-scale separation and identification for a large number of proteins, as the key experimental techniques in proteomics: 2-dementional electrophoresis (2DE) combined

Experimental Manual in Medical Biochemistry

with mass spectrometry, which allows rapid high-throughput identification of proteins and sequencing of peptides; Protein microarrays, which allow the detection of the relative levels of a large number of proteins present in a cell, and two-hybrid screening, which allows the systematic exploration of protein-protein interactions.

# Chapter V  Isolation and Purification of Protein

## Experiment 13    Isolation and Identification of Serum IgG

Immunoglobulin G (IgG) is the most abundant immunoglobulin and is approximately equally distributed in blood and in tissue liquids, constituting more than 70% of serum immunoglobulins in humans. Numerous approaches are well known for immunoglobulin isolation and purification. Here, a scheme for isolation and purification of IgG is described. Ammonium sulfate fractionation is the first procedure, followed by gel filtration for desalting, and DEAE-ion exchange chromatography is applied for further purification of IgG. Then SDS-PAGE and immunoelectrophoresis are used for identification of IgG in this experiment.

## Part I    Isolation and purification of IgG

### Principle

Salting-out is a process that protein precipitates at high salt concentrations. It can be used to fractionate proteins because the salt concentration of protein precipitation differs from one protein to another. Ammonium sulfate fractionation normally is the first purification step, and it can isolate nearly all of crude immunoglobulins from serum. The semi-saturated ammonium sulfate can precipitate globulins, while albumin is soluble. Then 33% saturated ammonium sulfate is used to precipitate out IgG from globulins. After several times of salting out, yielding about 40% pure preparation, enriched IgG is also very stable and suitable for long-term storage.

For further purification, IgG which mixed with ammonium sulfate should be desalted by dialysis or gel filtration. Dialysis is somewhat time-consuming. Gel filtration is a simple, gentle method for separating salt molecules and large proteins on basis of their size. Sephadex G-25 or G-50, as a kind of chromatographic matrix of porous beads, has been widely used. Smaller molecules enter into the pores of the beads and move through the column more slowly, while proteins enter less or not at all and thus move through more quickly.

Coupled with ion exchange chromatography (IEC), IgG will be purified from other contaminating proteins based on differences between the surface charges of the proteins. The ionic interaction occurs between the charge protein and an oppositely ion exchangers. IgG has a

Experimental Manual in Medical Biochemistry

higher or more basic pI than most serum proteins, it is positively charged in the pH 7.4 PBS buffer, while DEAE-cellulose is a kind of anion exchangers, therefore, IgG does not bind to column and directly elute from the IEC column. Finally, IgG maybe up the level of purification to more than 100-fold, and the yield normally is about 50% in this scheme.

## Reagents

1. Saturated ammonium sulfate solution
   Dissolve 800~950 g $(NH_4)_2SO_4$ in 1 L of hot distilled water ($dH_2O$), cool to room temperature and adjust pH to 7.2, store at 4℃.
2. 0.9% NaCl.
3. Sephadex G-25 or G-50.
4. 0.5 mol/L NaOH.
5. 0.01 mol/L phosphate buffer saline (PBS, pH 7.4).
6. 0.01 mol/L phosphate buffer saline containing 2 mol/L NaCl, pH 7.4.
7. DEAE-cellulose.
8. 0.5 mol/L HCl.
9. 1% $BaCl_2$.

## Procedure

### 1. Salting-out

(1) Put 10 ml serum and 10 ml 0.9% NaCl into a 100 ml beaker, and mix them thoroughly.

(2) Add saturated ammonium sulfate solution 20 ml dropwise to produce 50% saturation. Keep stirring on the magnetic stirrer 30 min to precipitate immunoglobulin completely.

(3) Centrifuge at 3000 r/min × 20 min. Discard the supernatant (mainly contains albumin). The precipitate contains all kinds of globulins.

(4) Dissolve the precipitate in with 10 ml of 0.9% NaCl and transfer to a 50 ml beaker.

(5) Add saturated ammonium sulfate solution 5 ml dropwise to produce 33% saturation. Go on stirring 30 min for precipitation of IgG.

(6) Centrifuge at 3000 r/min × 20 min. Discard the supernatant (mainly contains α-and β-globulin). IgG mainly exists in the precipitate.

(7) Repeat the above steps 3 times to remove coprecipitation (It will be contaminated with other proteins).

(8) Wash the precipitate with 33% saturated ammonium sulfate solution 3 times.

### 2. Desalting IgG—Gel Filtration

(1) Preparation of gel: Place about 15 g Sephadex G-25 or G-50 into a 1 L beaker, add 500 ml $dH_2O$ for 24 h, or boil gently for 2 h. After cooling, discard the supernatant and floating small particles, repeat washing 3~4 times.

(2) Packing column: Pour the above treated gel into a 1.0 cm × 50 cm glass column. The gel mixture will deposit and distribute naturally.

(3) Equilibration of column: Attach the constant flow pump and adaptor, and make the flow rate of $dH_2O$ be 2 ml/min at least. After 30 min, the column is equilibrated and ready for use.

(4) Connecting the fraction collectors.

(5) Loading sample: Completely dissolved precipitate with 1ml 0.9% NaCl, then load sample onto the column using a dropper, when the sample fluid just enter into gel surface, add $dH_2O$ onto the column (Note: Do not disturb the surface).

(6) Collection and detection: Elute protein using $dH_2O$ about 2ml/min flow rate, collect 2ml/tube fractions throughout this step. The automatic detector will detect white IgG flow out gradually and then mix them together. Later, $BaCl_2$ solution is used to check the elution of ammonium sulfate.

(7) Regeneration of column: Elute the column with $dH_2O$. When the elution peak returned to baseline and no ammonium sulfate in elution buffer detected by $BaCl_2$, the column can be used again.

### 3. DEAE-cellulose Ion Exchange Chromatography

(1) Preparation and packing column: Soak 15g DEAE-cellulose with 250 ml 0.5 mol/L HCl for 40 ~ 50 min, wash it to pH 4 with $dH_2O$. Then rinse in 250 ml 0.5 mol/L NaOH 50 min and wash to pH 8.0 with $dH_2O$. Finally equilibrate the column with 0.01 mol/L PBS (pH 7.4) and pack it into a 1.0 cm × 50 cm glass column. Attach the constant flow pump and adaptor.

(2) Loading and collection: load 1 ml desalted IgG solution, elute the column with 0.01 mol/L PBS using 1ml/min flow rate. Collect 4 ml every fraction tubes. The first protein peak is purified IgG.

(3) Regeneration of DEAE-cellulose column: pass through 2~3 column volumes PBS containing 2 mol/L NaCl, and wash thoroughly in PBS until $A_{280}$ of the effluent is less than 0.01, then add preservative and store at 4℃ for the next use.

## Part II  Identification of IgG product

### Principle

After using a purification procedure it is necessary to check the purity of IgG product and identify its specificity of the immune response. One of simplest methods for assessing purity of IgG fraction is by SDS-PAGE under non-reducing or reducing conditions. If proteins are running in SDS-PAGE without DTT pretreatment, the gel electrophoresis will separate native proteins according to difference in their charge density, and pure IgG will show one band on the gel after staining. If samples are electrophoresed under reducing conditions (adding DTT), staining IgG consists predominantly of two bands comprising heavy (~50 kU) and light (~22 kU) chains without other contaminating proteins.

Immunoelectrophoresis (IE) that combines electrophoresis and immunodiffusion, is a binary method of determining the levels of three major immunoglobulins: IgM, IgG, IgA. The sera proteins in a well are first separated by horizontal agarose gel electrophoresis on the basis of their different charge-to-mass ratios. IgG products are introduced into narrow troughs parallel to the separated sera antigens. Diffusion of both antigen and antibody takes place, and at a specific ration of antigen and antibody, a precipitin arc of antigen-antibody binding is formed on the plate, indicating the presence of specific IgG. The size, location, and shape of the precipitin arc, as compared to the control, are indications of the amount of protein in the test sample.

## Reagents

1. Acrylamide stock solution: 29 g acrylamide and 1.0 g bisacrylamide dissolve in 100 ml $dH_2O$ and through 0.45 μm filter. Stable at 4℃ for months. (Caution: it is a neurotoxin)
2. Separating gel buffer: 1.5 mol/L Tris-Cl, pH 8.8.
3. Stacking gel buffer: 0.5 mol/L Tris-Cl, pH 6.8.
4. 10% (w/v) Sodium dodecyl sulfate (SDS).
5. 10% Ammonium persulfate (AP) (store at 4℃ 1~2 weeks).
6. 10% N,N,N',N'-tetramethylethylenediamine (TEMED).
7. distilled water ($dH_2O$).
8. 2 × Sample loading buffer: Contain 100 mmol/L Tris-Cl, pH 6.8; 20% glycerol; 4% SDS; 0.02% Bromphenol blue (BB); 100 mmol/L DTT (nonreducing buffer without DTT).
9. Electrophoresis buffer: weigh out 3.0 g of Tris base, 14.4 g of glycine and 1 g SDS to make 1 L. The final pH should be 8.3.
10. Solutions for CBB staining and destaining:
    40% Methanol
    10% Acetic acid
    Staining solution contains 0.1% Coomassie Brilliant Blue R-250.
11. Protein markers.
12. Purified IgG or column fraction samples; serum as a control.
13. 1% agarose gel.
14. 5 × Tris-ethylbarbital electrophoresis buffer: 2.24 g ethylbarbital acid, 4.43 g Tris,

Chapter V  Isolation and Purification of Protein

0.053 g $Ca^{2+}$ saline of lactate, 0.065 g NaN3, in 100 ml $dH_2O$.

## Procedure

### 1. SDS-Polyacrylamide Gel Electrophoresis (PAGE) (see Experiment 5 in detail)

Table 5-3    Recipes for polyacrylamide separating and stacking gels

| Reagents | 3% Stacking Gel (ml) | 7.5% Separation Gel (ml) |
| --- | --- | --- |
| Acrylamide stock solution | 1.0 | 2.5 |
| Separating gel buffer | – | 2.5 |
| Stacking gel buffer | 1.25 | – |
| $dH_2O$ | 2.6 | 4.75 |
| 10% SDS | 0.05 | 0.1 |
| 10% AP | 0.05 | 0.1 |
| 10% TEMED | 0.05 | 0.05 |
| Total volume | 5 | 10 |

(1) Preparing a vertical slab Polyacrylamide gel (7.5% separation gel and 3% stacking gel) as described in table 5-3.

(2) Preparing and loading Samples: Determine IgG concentration by Bradford method or calculate by $A_{280}$ (a 1mg/ml IgG solution will have an $A_{280}$ of 13.6). Dilute a portion of IgG sample with 2 × SDS sample loading buffer and heat for 3~5 min at 100℃. Load approx 10~15 μg IgG per well.

(3) Running the gel: Assemble gel cell and connect the electrodes to a power pack, turn on power supply to run the gel using 10 mA of constant current at start, 30 mA at end for a 1.0 mm gel until BB dye reaches the bottom of the gel. Turn off the power.

(4) Proteins detection by staining: Remove gel from plates carefully, and stain with CBB staining solution for 1 h (gently rocking) or preferably overnight. Then the gel is destained with destaining solution until blue bands and a clear background are obtained.

### 2. Immunoelectrophoresis (IE) of IgG product

(1) Preparation of agarose gel: Mix agarose powder with electrophoresis buffer and heat in 90℃ water bath until dissolve to make 1% agarose gel. Spread gel in horizontal glass plate and let the gel to cool, puncture wells with 4 mm diameter drilling equipment and make the trough with razor according the figure 5-3.

(2) Running electrophoresis: Add 4%~5% serum, which is diluted with electrophoresis buffer containing 0.01% bromophenol blue (BB) into the well and put the gel plate in

the electrophoresis chamber. Communicate two ends of plate to the buffer with wetted filter paper and put the cover on the chamber. Turn on power supply, run the gel at the 6 V/cm for 1hr with a migration distance of 35 mm (when BB dye reaches the end of the gel).

(3) Diffusion: Remove the agar plate from the chamber and put it on a flat surface. Remove the buffer in trough and add appropriate amount of IgG (0.1~0.2 ml) into the trough. Incubate the gel plate at 30℃ in a humidity chamber containing a moist paper wick for at least 12 h. Transfer the plate to 0.85% saline, which can stop the diffusion and wash out unbound proteins, then a precipitin arc of antigen-IgG binding is visible on the plate by naked eyes or staining with CBB dye after drying the gel.

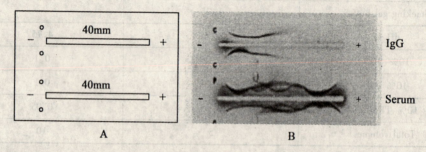

Fig. 5-3 Preparation of agarose gel plate and result of immunoelectrophoresis.

## Chapter V  Isolation and Purification of Protein

**Experiment title**

**Date**

**Observations and results**

(Calculate the yield of IgG product after IEC. Check the purity of IgG by SDS-PAGE and observe the precipitation arc of IgG product which is formed in immunoelectrophoresis)

## Discussion

1. Describe briefly the main methods for separation and purification of proteins?
2. Describe the methods for identification of purified protein.
3. The most general method to monitor the purification process is by running a —— in the different steps.

**Teacher's remarks**

**Signature**

**Date**

Experimental Manual in Medical Biochemistry

## Experiment 14  Expression, Purification and Identification of Glutathione S-Transferase Fusion Protein

Fusion proteins are commonly used for producing foreign proteins in *Escherichia coli* (*E coli*). The glutathione S-transferase (GST) fusion system using pGEX expression vector has proven successful in producing properly folded and biologically active proteins or polypeptides. Each pGEX vector contains an open-reading frame encoding GST, followed by unique restriction endonuclease site (*Bam*H I, *Sma* I, *Eco*R I) and termination codons. If a target gene is inserted into pGEX-2T or pGEX-3X vector at the end of the GST gene, and transferred to *E coli*, GST fusion protein will be high-level expressed on induction with isopropy-β-D-thiogalactoside (IPTG). Using some convenient methods for specific purification of GST fusion protein from cellular contaminants, and the site-specific protease (such as thrombin) enzymatic cleavage of GST protein, the target proteins are readily amenable to the study of their biological activities and/or interactions.

## Part I  Expression and Purification of GST Fusion Protein

### Principle

GST regions often can be exploited in purifying the target protein. With the use of glutathione as a specific ligand for the GST protein, affinity precipitation is a simple and the most powerful for purification of GST fusion protein. Because glutathione is covalently coupled on a solid support (Sepharose, agarose or certain water-soluble polymers), combining the affinity interaction and the precipitation, GST-glutathione complex can precipitate and separate from other components in solution. Then, GST/fusion protein can elute by washing with elution buffer containing excess glutathione, or the target proteins are obtained by enzymatic or chemical cleavage from GST fusion protein.

### Reagents

1. pGEX vector.
2. Competent cells of *Escherichia coli* (*E coli*).

Chapter V Isolation and Purification of Protein

3. LB plates containing 50 μg/ml ampicillin (Amp).
4. LB medium containing 50 μg/ml ampicillin (Amp).
5. 1 mol/L isopropy-β-D-thiogalactoside (IPTG): Weigh 2.38 g IPTG and add ddH$_2$O to 10 ml, then filter with 0.2 μm of filter membrane, aliquot and store at -20℃.
6. 0.01 mol/L Sodium phosphate buffer (PBS, pH 7.4).
7. Suspension solution of glutathione agarose beads.
8. 10% Triton X-100.
9. GST wash buffer: 50 mmol/L Tris·Cl (pH 7.5) /150 mmol/L NaCl.
10. GST elution buffer: 50 mmol/L Tris-Cl (pH 8.0)/5 mmol/L reduced glutathione (GSH). (fresh prepare, pH 7.5).
11. Reagents for SDS-PAGE (see experiment 13 part II).

## Procedure

1. Subclone the chosen DNA fragment into pGEX vector, transform competent *E coli* and screen the colony containing transformant on LB/Amp plates, pick it up and inoculate it in 2 ml of LB/Amp medium. Include a control for setting the transformed bacteria with pGEX vector. Grow cultures with vigorous agitation in a shaking incubator at 37℃ until visibly turbid (3~5 h).
2. Induce fusion protein expression by adding 1 mol/L IPTG to a final concentration of 0.1 mmol/L. Continue shaking at 37℃ for another 3~6 h.
3. Transfer the liquid culture to a microcentrifuge tube. Centrifuge the culture 10 000g × 15sec, and discard supernatant. Resuspend pellets in 300 μl ice-cold PBS. Transfer 10 μl to another centrifuge tube for checking the expression of GST/fusion protein by SDS-PAGE.
4. Lyse cells by using a sonicator with a 2 mm probe and then centrifuge at 10 000g × 5 min at 4℃. Transfer the supernatant to a new centrifuge tube.
5. Add 50% 50 μl glutathione agarose beads suspension to supernatant and mix gently >2 min at room temperature. Add 1 ml PBS and spin at 10 000g × 5 sec. Collect beads and discard supernatant. Wash the beads 2 times with PBS.
6. Add an equal volume of 1 × SDS sample loading buffer to above beads, and 30 μl to the 10 μl resuspension cells (from step 3), heat 3 min at 100℃, centrifuge briefly. After loading, run the 10% SDS-PAGE. GST protein from pGEX and fusion protein can be visualized by staining with Coomassie brilliant blue.
7. Inoculate a colony of the pGEX transformant into 50 ml LB/Amp medium, grow 12~15 h at 37℃ in a shaking incubator.
8. Dilute above culture 1:10 into 500 fresh LB/Amp medium, culture at 37℃ for 1 h, add 1 mol/L IPTG to 0.1 mmol/L (final) and keep on cultivating for 3~7 h.
9. Centrifuge cells 10 min at 5000g, discard supernatant and resuspend the pellet with 10~20

ml ice-cold PBS.

10. Lyse cells by sonicator with 5 mm diameter probe within 30 sec, adjust frequency and intensity of sonication so lysis occurs in ~30 sec, without frothing.

11. Add 10% Triton X-100 to 1% and mix. Centrifuge cells at 10000g × 5 min at 4℃, to remove insoluble materials.

12. Collect supernatant carefully. Add 1 ml 50% glutathione agarose beads suspension and mix gently over 2 min at room temperature. Wash by adding 50 ml ice-cold PBS and repeat two more times. Resuspend beads with ice-cold PBS in a small volume (1 ~ 2 ml) and transfer to an Eppendorf tube.

13. Discard supernatant by short centrifuge and collect the beads. Elute fusion protein by adding 1 ml 50 mmol/L Tris-Cl (pH 8.0)/5 mmol/L reduced glutathione. Mix gently for 2 min. Centrifuge cells at 500g × 10 sec and then collect supernatant. Repeat the wash 2 ~ 3 times and analyze each fraction by SDS-PAGE. Store eluted protein in aliquots containing 10% Glycerol at -70℃. Determine the yield of fusion protein by measuring $A_{280}$. For GST protein, $A_{280} = 1$ corresponds to a protein concentration of 0.5 mg/ml.

## Part II  Detection and Identification of GST Fusion Protein

### Principle

Separation of protein by SDS-PAGE normally is one of simplest methods for detecting purity of protein product. In a denatured state, with strong reducing agents to remove secondary and tertiary structure, proteins are separated according to size. Hence, the purity of protein sample can be identified by the number of bands on the gel after staining.

Western blotting (alternately, immunoblotting) that combines electrophoresis and protein blotting, is a method of detecting specific protein in a crude extract or a more purified preparation. The first step uses gel electrophoresis (normally PAGE or SDS-PAGE) to separate native or denatured proteins. In order to make the proteins accessible to antibody detection, the proteins within the gel are then electrotransferred onto a membrane made of nitrocellulose (NC) or polyvinylidene fluoride (PVDF). Non-specific protein binding on the membrane is based upon hydrophobic interactions, as well as charged interactions. So after blocking non-specific binding on the membrane using bovine serum albumin (BSA) or non-fat milk, with a detergent such as Tween 20, antibodies (both monoclonal and polyclonal antibodies) only attach on the binding sites of the specific target protein. Last detection for the protein of interest is "probed" by a modified antibody which is linked to a reporter enzyme, such as alkaline phosphatase (AP), or horseradish peroxidase (HRP), which drives a colorimetric reaction or chemiluminescence reaction (Figure 5-4).

Western blotting also allows investigators to measure relative amounts of the protein present in different samples. Normally the process is repeated for a structural protein, such as β-actin or tubulin, which should not change between samples, to control between groups.

## Reagents

1. Reagents for SDS-PAGE (see experiment 13 part II).
2. Prestaining protein standards.
3. 1 × Transblotting buffer: Dissolve 3.03 g Tris base; 14.4 g Glycine into 200 ml $dH_2O$, mix with 200 ml Methanol, add $dH_2O$ up to 1 L. pH should be 8.3 without adjustment.
4. TBS buffer: Add 1.22 g Tris (10 mmol/L) and 8.78 g NaCl (150 mmol/L) to 1L $dH_2O$, and adjust pH to 7.5 with HCl.
5. TBS-T buffer: 1L TBS buffer add 0.5 ml Tween 20 (0.05%).

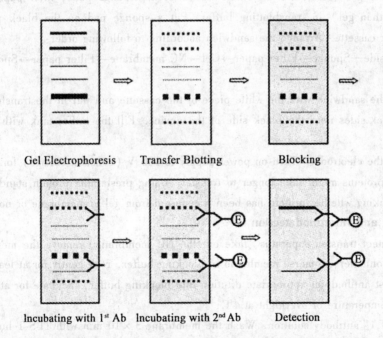

Fig. 5-4 The basic procedure of Western blotting

6. Blocking buffer: Dissolve 3.0 g Bovine serum albumin (BSA) or 5.0 g non-fat milk into 100 ml TBS-T buffer, keep at 4℃.
7. First (primary) antibody (1st Ab): rabbit anti-GST monoclonal Ab.
8. Second antibody (2nd Ab): goat anti-rabbit IgG (HRP labeled).
9. Developing reagent: DAB 6 mg dissolve in 10 ml TBS-T buffer, and mix with 0.05 ml 30% $H_2O_2$. Enhanced chemiluminescent (ECL) kit can also be used.

## Procedure

### 1. Seperation of protein by SDS-PAGE

(1) Preparation of samples:

① Cell lysates: Collect cells and lyse the cells pellet with 100 μl SDS-PAGE sample loading buffer. Boil for 5 min, then cool at room temperature.

② Protein solution: After determine the protein concentration (Bradford assay, $A_{280}$, or BCA method), mix with equal volume 2 × sample loading buffer, boil for 5 min and cool.

(2) Loading and running the gel (see experiment 5): After separation, take out the gel, cutting a corner of gel for orientation and put gel in the transblotting buffer.

### 2. Electroblotting

(1) Preparation of membrane: Cut a piece of NC membrane according to the size of gel, and wet for about 10 min in transblotting buffer.

(2) Assemble gel-membrane "sandwich": Prewet the sponge pads, filter papers (slightly bigger than gel) in transblotting buffer. Put a sponge pad on the black piece of the transfer cassette. Arrange the sandwich according to following order:

Black side—Sponge→Filter paper→Gel→NC membrane→Filter paper→Sponge—White side.

Close the sandwich with the white piece of the cassette and put in the transfer tank, put the black sides near the black side of the device. Fill the buffer tank with the transfer buffer.

(3) Attach the electrodes. Turn on power supply to 100 V (constant voltage) for 1 h at 4℃. Bigger proteins might take longer to transfer. Using prestaining protein standards is easy for checking whether protein has been transferred from gel to membrane or not.

### 3. Blocking and Immunodetection

(1) Disconnect transfer apparatus, take out the NC membrane, remove the membrane to a small container, immerse membrane in blocking buffer, rock gently for at least 1 h.

(2) Add first antibody at appropriate dilution into blocking buffer, incubate for at least 1 h at room temperature or overnight at 4℃.

(3) Pour off 1$^{st}$ antibody solution. Wash the membrane 3 × 10 min with TBS-T buffer.

(4) Incubate with secondary antibody diluted in blocking buffer for at least 1 h.

(5) Pour off 2$^{nd}$ antibody solution. Wash the membrane 3 × 10 min with TBS-T buffer.

(6) Detected the specific protein band with enhanced chemiluminescent (ECL) kit or detected with colorimetric reaction: Pour off TBS-T buffer, cover developing reagent to the membrane, monitoring development.

## Notes

1. Acrylamide and bisacrylamide are neurotoxic and may be absorbed through the skin.

Chapter V  Isolation and Purification of Protein

Polyacrylamide is considered nontoxic but may contain unpolymerized material. When preparing and handling gels, you should wear gloves to avoid exposure to unpolymerized polyacrylamide.

2. One major difference between NC and PVDF membranes relates to the ability of each to support "stripping" off and reusing for subsequent antibody probes. The sturdier PVDF allows for easier stripping, and for more reuse before background noise limits experiments. Another difference is that PVDF must be soaked in 95% ethanol or methanol before use. PVDF membranes also tend to be thicker and more resistant to damage during use.

**Experiment title**
**Date**
**Observations and results**

## Discussion

1. GST protein as a tag-protein is often used to expression and purification of foreign protein. With the use of _____ as a specific ligand for the GST protein, _____ is a simple and the most powerful for purification of GST fusion protein.
2. What is the principle and application of Western blotting? Why need "blocking" the NC membrane after electroblotting?
3. Describe the difference between NC and PVDF membrane.

**Teacher's remarks**
**Signature**
**Date**

# Chapter VI
# Isolation, Purification and Identification of Nucleic Acids

## 6.1 Isolation and Purification of Nucleic Acids

Isolation and purification of nucleic acids from various sources is the most important first step in molecular biology research, and the quality of nucleic acids product will be directly related to the success or failure of the further experiment.

Nucleic acids including DNA and RNA are present as nucleoprotein complexes in cells. Eukaryotic chromosomal DNA exists as a double-stranded helical molecule. "Chromosome", plasmid DNA in prokaryotes and mitochondria DNA in eukaryotic cells are double-stranded circular molecules. RNA exists in several different single-stranded structures with different characteristics, such as the poly A tail of eukaryotic mRNA. 95% of the eukaryotic DNA is in the nucleus, others are organelle DNA, such as mitochondria DNA, etc. 75% of RNA exists in the cytoplasm, and 10% in the nucleus. Others are in the organelle.

The criteria of isolation and purification of nucleic acids are as the following: ① Maintaining the integrity of primary structure, which is essential for further research of nucleic acids; ② Removing contaminating materials such as proteins, lipids, polysaccharides and other high molecular weight compounds, including removing other types of nucleic acids molecules, for example, the contaminating DNA needs to be removed when isolating RNA.

To ensure the integrity and purity, the preferred method for isolation of nucleic acids from any source should be simple and rapid. Usually, it have been considered the following problems: ① Reduce chemical degradation of nucleic acids, avoiding the cleavage of phosphodiester bond by base and acid, solution of pH 4~10 is common selective; ② Reduce mechanical damage for high molecular weight nucleic acids, such as vortex, strong stirring, repeated freezing-thawing, etc. Small circular plasmid DNA and RNA are relatively resistant to mechanical damage; ③ Prevent hydrolysis of nucleases for nucleic acids. Adding EDTA and citrate to buffer can chelate $Mg^{2+}$, $Ca^{2+}$ ions which are required for DNase activity. Especially in the process of RNA isolation, because of high activity of DNase, cell lysis buffer should contain RNase inhibitor such as bentonite, guanidine isothiocyanate, selecting fresh tissue or cell samples will be the best choice. Materials should be stored in liquid nitrogen or -70℃ if not used immediately.

The main steps involved in isolation of nucleic acids include cells disruption and removal of contamination of another DNA or RNA, protein, polysaccharides, lipids, other macromolecules. At last, salts and organic solvents should be discarded if need purified nucleic acid molecules.

### 6.1.1 Isolation of eukaryotic genomic DNA

DNA is very inert molecules with double strands closely linked by hydrogen bonds. In addition, bases are protected by lateral phosphate and ribose, which strengthened the internal stacking force. Nevertheless, high molecular weight DNA is still fragile. Long and curved DNA molecules in the solution are easy to break by mechanical damage such as fluid suction, vortex, and strong stirring. Therefore, it is difficult to obtain larger molecular weight DNA, especially for more than 150 kb.

All nucleated cells in eukaryotes (including culture cells) can be used for genomic DNA preparation. The isolation protocols consist of two parts: a technique to lyse gently the cells and solubilize the DNA, followed by several enzymatic or chemical methods to remove contaminating RNA, proteins, polysaccharides, and other macromolecules.

Various methods used to break eukaryotic cells include ultrasonication, homogenation, and low osmotic pressure etc. Gentle detergent or proteinase K treatment is usually used to obtain high molecular weight DNA.

Proteins are removed by treatment with phenol or mixture of chloroform-isoamyl alcohol or phenol/chloroform. Repeated extraction operation may damage the DNA molecules by shearing force. So using 80% formamide plus dialysis method can obtain DNA fragments up to 200 kb. It depends on different requirements to select different isolation methods.

### 6.1.2 Isolation of plasmid DNA

A plasmid is an extra-chromosomal, circular DNA molecule that replicates independently of the chromosomal DNA of bacteria. It is much smaller than chromosomal DNA (MW range from 1 kb to 200 kb). The replication of plasmid includes two types: stringent control and relaxed control. The former is low copy number plasmids. The latter is high copy number plasmids (usually contains more than >100 copies).

Plasmids can carry foreign DNA into bacteria for amplification or expression, so it is called vectors in genetic engineering. Many plasmids are commercially available (i.e. pUC18, pBR322 etc.). A typical plasmid (vectors) used must contain an origin of replication (ori), a region called multiple clone site (MCS) for the insertion of the experimental DNA, and contain a selectable marker such as a gene for resistance to an antibiotic (e.g. ampicillin).

The isolation and extraction of plasmid DNA are the most popular techniques in molecular cloning. The isolation of plasmid normally contains three steps: growth of bacteria containing the plasmid of interest, separation and lysis of the bacteria from the culture medium, and purification of the plasmid DNA.

Methods for preparation of plasmid DNA include SDS-alkaline denaturation, salt-SDS precipitation and rapid boiling, etc. The SDS-alkaline denaturation method is a popular procedure for plasmid minipreparation because of its overall versatility and consistency. The gradients centrifugation in CsCl-ethidium bromide is useful for large-scale preparation of macro, closed plasmid DNA. But it is expensive and time-consuming, and has been substituted by ion exchange chromatography, gel filtration now. Plasmid purification kits including spin chromatographic column are also common used, and are relatively expensive. Plasmid DNA purified by above-mentioned methods can be used for PCR, bacteria and mammalian cell transformation, restriction enzyme analysis, regular subclone, and probe labeling.

### 6.1.3 Isolation of tissue/cell RNA

Isolation of pure and intact RNA from cells is crucial for many molecular biological experiments, and is the basis of gene expression analysis. For example, the successful performing of Northern blotting, cDNA synthesis and in vitro translation experiments is extremely determined by the quality of RNA.

A typical mammalian cell contains about $10^{-5}$ μg RNA, of which 80% ~ 85% is rRNA (mainly including 28S, 18S, 5.8S and 5S), and only 1% ~ 5% is mRNA. High abundant RNA is highly homologous each other. According to their density and molecular size, separation methods involve density gradient centrifugation, gel electrophoresis, anion-exchange chromatography and high-performance liquid chromatography (HPLC). Various mRNA which function as messengers transferring genetic information, is one of the main research subjects in molecular biology. Eukaryotic mRNA molecules have a wide range of molecular sizes and different nucleotide sequences, but with 3'-polyA tail (20 ~ 250 nucleotides in length), therefore, mRNA can be separated using oligo (dT)-cellulose chromatography.

For a wide range of needs and applications in molecular biology, many simple and high-yielding methods have been devised recently for isolation and purification of total RNA, mRNA and specific RNA.

With 2', 3' hydroxyl residues of ribose, RNA has more active than DNA. It is high sensitive to RNA enzyme (RNase), which has very stable biological activity, particularly pancreatic RNase. RNase exists not only in cells, but also in dust, various containers and test reagents, human sweat and saliva. It is heat, acid and alkali resistant, and its activity does not influenced by divalent metal ion chelators. Therefore, the key for RNA isolation is controlling the contaminating RNase, including exogenous RNase and endogenous RNase.

(1) Prevention of contamination with exogenous RNase

Exogenous RNase mainly comes from operators' hands, saliva, and cells, mycetes in the dust. Therefore, disposable masks and gloves should be worn, and clean environment is also necessary. All operations should be conducted on ice, as low temperature can reduce the activity of RNase.

Wash all glasswares and roast at 180℃ for 3 h before use, or rinse all glasswares with 0.1% diethylpyrocarbonate (DEPC) water to inactivate RNase and then autoclave them to evaporate DEPC. Choose disposable RNase-free plastics (such as the Eppendorf tubes, tips), or treat with 0.1% DEPC water and autoclave them before use. Electrophoresis tanks should be soaked in 3% $H_2O_2$ for 10 min, then rinse with DEPC treated water thoroughly. It is better to prepare dedicated tank for RNA electrophoresis.

All solutions should be added DEPC to 0.05% ~ 0.1%, put in room temperature overnight, or stirred by magnetic force in room temperature for 20 min before autoclave (15 Ibf/in$^2$, 15 ~ 30 min). Some solutions cannot be autoclaved. They should be made using sterilized DEPC-treated dd$H_2O$, and sterilized through 0.22 μm membrane filter.

DEPC will react with Tris base rapidly (only 1.25 min of half-life), and decompose into $CO_2$ and ethanol. Tris buffers should be made up of sterilized DEPC-treated dd$H_2O$ and Tris Crystal unopened. Phenol for RNA extracting protein during RNA isolation should be prepared separately, and added 8-hydroxyquinoline to 0.1%, which can inhibit RNase and oxidant.

(2) Inhibiting the activity of endogenous RNase

Release of endogenous RNase from tissue or cell is the major reason to result in degradation of RNA when lysing cells. It is very important to inhibit the RNase activity as early as possible by removing cell protein and adding RNase inhibitor such as guanidine isothiocyanate.

The amount of endogenous RNase differs in tissues. Pancreas and spleen cells are rich in RNase, and it is essential to use several strong RNase inhibitors.

## 6.2 Identification and Analysis of Nucleic Acids

The application of molecular biology techniques to the analysis of complex genomes depends on the ability to prepare high-MW DNA and RNA with high quality. Pure and intact nucleic acids are the basic requirement for further research of structure and function. Commonly used methods for identification and analysis of nucleic acids include UV spectrophotometry, gel electrophesis, chromatography, hybridization, PCR and DNA sequencing, etc.

# Chapter VI  Isolation, Purification and Identification of Nucleic Acids

## 6.2.1  Spectrophotometry

Purine and pyrimidine bases of nucleic acid molecules both have UV absorption characteristics. The maximum absorption peak is at 260 nm wavelength, which can be used for quantitative analysis of nucleic acids. UV spectrophotometry is only used to determine the concentration of nucleic acid solutions more than 0.25 μg/μl. For a quite diluted solution, fluorescence spectrometry can be used to estimate the concentration of nucleic acids. The absorption of 1 OD ($A$) is equivalent to approximately 50 μg/ml double-stranded DNA (dsDNA), 33 μg/ml single stranded DNA (ssDNA), or approximately 40 μg/ml RNA. The ratio $A_{260}/A_{280}$ is used to estimate the purity of nucleic acid, since proteins absorb at 280 nm. Pure DNA should have a ratio of approximately 1.8, whereas pure RNA should give a value of approximately 2.0. Absorption at 230 nm reflects contamination of the sample by substances such as carbohydrates, peptides, phenols or aromatic compounds. In the case of pure samples, the ratio $A_{260}/A_{230}$ should be approximately 2.2.

## 6.2.2  Gel electrophoresis of nucleic acids

Gel electrophoresis separations can be used either for preparation or for analysis of nucleic acids. Specific molecular weight DNA fragments can be recovered and purified after separation by gel electrophoresis. Nucleic acids are uniformly negatively charged, and will migrate to the anode in electric field. Nucleic acid molecules have similar charge density. Choosing appropriate concentration of gel as electrophoresis support medium, which has a molecular sieve effect, can separate nucleic acid molecules based on different molecular size and conformation.

### 6.2.2.1  Factors affecting DNA electrophoresis

(1) The nature of nucleic acids

Charge, molecular size and conformation of nucleic acids determine the electrophoretic mobility. Generally, linear dsDNA molecules don't have complex conformation affecting the mobility, and the relationship between relative mobility and common logarithm of molecular weight is inversely proportional in the gel electrophoresis. Molecular conformation can also affect the mobility. For example, the mobility of plasmid DNA with the same molecular weight is: closed-loop > linear > open-loop. The mobility of ssDNA or RNA is affected by base composition and sequence in gel electrophoresis. Therefore, single-stranded nucleic acids can be separated using denaturing gel electrophoresis.

(2) Pore size of gel

Agarose and polyacrylamide gel (PAG) is support materials often used for nucleic acid gel electrophoresis. Changes of gel concentration can adjust the pore size of the gel, which dictates the size of the fragments that can be resolved. Agarose gels have larger pore size and lower resolution, but have a wide range to resolve DNA fragments from 100 bp to 50 kb. DNA

Experimental Manual in Medical Biochemistry

fragments less than 20 kb is mostly suitable to horizontal gel electrophoresis separation in the electric field with constant intensity and direction. Larger DNA molecules require some special pulsed-field gel electrophoresis (PFGE) to be separated. Polyacrylamide gel can be used to separate DNA fragments from 5 bp to 500 bp, and can even tell the difference of 1 bp between DNA fragments. Table 6-1 lists different gel concentration with their corresponding separation range of linear DNA.

Table 6-1    **Gel concentration and effective separation range of linear DNA**

| Gel | Gel concentration(%) | Separation range of linear DNA (bp) | Bromphenol blue (bp) (equivalent to DNA) |
|---|---|---|---|
| Agarose | 0.3 | 5000 ~ 60000 | |
| | 0.7 | 800 ~ 10000 | 1000 |
| | 0.9 | 500 ~ 7000 | 600 |
| | 1.2 | 400 ~ 6000 | |
| | 1.5 | 200 ~ 4000 | |
| | 2.0 | 100 ~ 3000 | 150 |
| PAGE | 3.5 | 100 ~ 2000 | 100 |
| | 5.0 | 80 ~ 500 | 65 |
| | 8.0 | 60 ~ 400 | 45 |
| | 12.0 | 40 ~ 200 | 20 |
| | 15.0 | 25 ~ 150 | 15 |
| | 20.0 | 6 ~ 100 | 12 |

(3) Electric field strength and direction of the electric field

The mobility of linear DNA fragments is directly proportional to the voltage when voltage is low. Usually the electric strength is not more than 5 V/cm gel. If electric field intensity increases, migration of high molecular weight DNA fragments will be increased in different range, and the effective separation range of the gel will reduce, 0.5 ~ 1.0 V/cm overnight electrophoresis is often used to separate high molecular weight DNA fragments in order to get higher resolution and neat bands.

(4) Electrophoretic buffer

Buffer composition and ionic strength directly affect electrophoretic mobility. Gels are usually run in Tris-acetate-EDTA (TAE), Tris-borate-EDTA (TBE) or Tris-phosphate-EDTA (TPE) buffer. TAE is used traditionally, but its buffer capacity is very low. TBE and TPE have higher buffer capacity, but borate can form complexes with agarose in TBE buffer. So in a

# Chapter VI  Isolation, Purification and Identification of Nucleic Acids

short period of agarose gel electrophoresis (≤5 h), TAE is still used, and TBE is often used in PAGE.

### 6.2.2.2  Tracker and dye in gel electrophoresis of nucleic acids

(1) Tracker

Colored markers are usually used in the sample buffer during electrophoresis to indicate the migration of samples, which include bromphenol blue (BB) and green xylene (blue). Molecular weight of BB is 670 U. The mobility of BB is almost equivalent to 1 kb and 0.6 kb, 0.15 kb of linear dsDNA fragments in 0.6%, 1%, 2% agarose gel electrophoresis, respectively. The molecular weight of green xylene is 554.6 U, and the quantity of electricity is less than BB, which results in slower mobility. The mobility of green xylene is equivalent to 4 kb linear dsDNA fragment in 1.0% agarose gel electrophoresis.

(2) Dye

After electrophoresis, nucleic acids need to be stained to show the bands. The following are dyes for nucleic acids usually used:

① Ethidium bromide (EB)

EB is the most common fluorescence dye for nucleic acids, which can embed into double-stranded DNA base pairs, and issue orange fluorescence in the ultraviolet excitation. The fluorescence intensity of EB-DNA complexes is 10 times stronger than free EB in gels. By this method, 10 ng of DNA samples can even be detected. EB can also be used to detect ssDNA/RNA, but the fluorescence yields are relatively low because of lower affinity with single-stranded nucleic acids.

Adding EB to a final concentration of 0.5 μg/ml in the gel can observe the migration of nucleic acids at any time. But with positive charge, EB will increase the length of open-loop and linear DNA molecules, and make them more rigid after embedding into bases. The mobility of linear DNA will drop 15 percent rate. Therefore, the gel should be immersed in 0.5 μg/ml of EB solution for 10 min after electrophoresis if it is used for measure the molecular weight of nucleic acid. EB is easily decomposed in the light, and should be kept dark at 4℃.

② $Ag^+$

$Ag^+$ can form stable complexes with nucleic acid, and then be reduced to silver particles by formaldehyde. $AgNO_3$ reagents make DNA and RNA dark brown in PAG. The sensitivity of silver staining is about 200-fold higher than EB staining. It can detect RNA of 0.5 ng in less than 0.5 mm thick gel. But $Ag^+$ can also bind with protein, detergent and make them brown. $Ag^+$ binds with DNA stably and damages DNA, so silver staining is not suitable for the recovery of DNA fragments.

### 6.2.3  Nucleic acids hybridization

Identification of some specified fragments from complex mixtures separated by gel electrophoresis usually uses nucleic acids hybridization (such as Southern blotting and Northern blotting analysis). First transfer all fragments to nylon membrane or nitrate cellulose membrane from the gel, and then use labeled nucleic acid probe to hybridize with target fragments for identification.

### 6.2.4 Polymerase chain reaction

Polymerase chain reaction (PCR) is a rapid procedure for in vitro enzymatic amplification of specific DNA fragments. DNA replication is carried out in tubes, and millions of copies of the specific DNA sequence can be obtained in a short period of time from biological material for gene amplification, separation, screening, sequence analysis or identification. PCR has revolutionized molecular genetics by permitting rapid cloning and analysis of DNA. Since the first reports describing this new technology in the mid 1980s, there have been numerous applications in molecular cloning, diagnosis of genetic diseases, forensic science, and archaeology etc.

#### 6.2.4.1 Principle of PCR

PCR is actually a DNA polymerase-dependent enzymatic synthesis reaction in the presence of a suitably heat-stable DNA polymerase, DNA precursors (the four deoxynucleoside triphosphates, dATP, dCTP, dGTP and dTTP), template DNA, and primer (complementary in sequence to a specific region of the template DNA), depending on the binding specificity of PCR primers and the template DNA.

The amplification reaction involves sequential cycles composed of three steps:

(1) **Denaturation**: template DNA is heated to denature the dsDNA molecules, making them single-stranded;

(2) **Dnnealing**: the reaction mix is then cooled, allowing primers to anneal to complementary sequences on opposite strands of the template DNA (by hydrogen bonding between complementary bases: A-T, G-C) flanking the DNA segment to be amplified;

(3) **Extension**: the reaction is then brought to an intermediate temperature, and, using free deoxyribonucleotides added to the reaction mixture, DNA polymerase extends these primers from their 3' ends toward each other. This three-step process is repeated for a number of cycles (usually 20 ~ 30 cycles), resulting in the production of many copies ($10^6 \sim 10^9$) of the template DNA sequence. The practical operation of PCR includes preparation of template DNA, primer design and synthesis, enzymatic polymerization reaction, and detection of the reaction product.

#### 6.2.4.2 Reaction system of PCR

(1) **Design principle of primer**

As an in vitro enzymatic reaction, the efficiency and specificity of PCR is based on two aspects: specific binding of primer and template, and the effective primer extension by

# Chapter VI  Isolation, Purification and Identification of Nucleic Acids

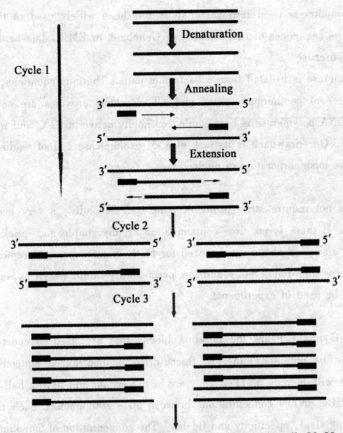

This three-step process is repeated for a number of cycles(usually 20~30 cycles), resulting in the production of many copies ($10^6$~$10^9$) of the template DNA sequence.

**Fig. 6-1  PCR principle**

polymerase. To reduce non-specific amplification products, the general design principle of primer is to raise the efficiency and specificity of PCR. Here are basic rules in designing primers: ① The length of 15 ~ 30 base is suitable. Too short or too long primers will have lower specificity. ② G + C content of primer should be 45% ~ 55%, with random distribution of the base. ③ It should not contain complementary sequences between or in primers, avoiding hairpin-like secondary structure or primer dimers. ④ Specificity: the homology between the base sequence of primer and non-amplified region should be less than 70 percent, or less than eight consecutive complementary bases. ⑤ 3' end of primer must be matched with the DNA template. Try to make it terminate in C or A, as neither nucleotide wobbles. It is better if the 3' end is slightly AT rich, as it will be less prone to misprime. ⑥ It has no strict limitations for 5' terminal of primer, which can be in a free state, not matching with the DNA template. Therefore, the primers can be designed with the restriction endonuclease sites or other short sequences (such as ATG initiation code or a mismatched base mutation, etc.) at the 5' end.

At present, computer-assisted retrieval analysis has been widely used in the design of primers, and the gene sequences can be found from Genebank or EMBL data-bank.

#### (2) Taq DNA polymerase

Taq DNA polymerase is isolated from a bacterium called Thermus aquaticus, which lives in the very hot water of geothermal vents, and all of whose enzymes are active at high temperatures. Taq DNA polymerase is heat stable, optimally active at 72℃ and work over the pH range 7.0 ~ 7.5. One drawback is its lack of 3'-5' exonuclease (proof reading) activity, which can lead to the misincorporation of nucleotides.

#### (3) Template

Most PCR does not require strict purity of template. Secondly, a very low amount of templates is enough. If there is no cross-contamination, many simple and quick DNA/RNA preparation methods discussed above can be used for PCR. Sometimes more rapid and simple methods, such as using high-temperature and low permeability liquid (such as water boiling) to dissolve cells can be used in experiments.

#### (4) dNTPs

dNTPs (A. T. G and C) are the building blocks that are used to construct genetic molecules. They can be obtained either as freeze-dried or neutralized aqueous solutions. Repeated freeze-thaw will degrade dNTPs. They are heat resistant and have a half-life of more than 40 cycles of PCR. dNTPs concentrations between 50 ~ 200 mmol/L each result in the optimal balance among yield, specificity and fidelity. The concentration of four kinds of dNTPs should be used at equivalent concentrations to minimize misincorporation errors.

#### (5) Reaction buffer

The recommend buffer for PCR is 10 mmol/L Tris buffer, with a pH range between 8.5 and 9.0 at 25℃. Because a buffer made to pH 8.8 at 25℃ will have a pH value of 7.4 at 72℃, this value is optimal for the activity of Taq polymerase at this temperature.

#### (6) Magnesium ions

The magnesium concentration may affect all of the following: primer annealing, strand dissociation, product specificity, formation of primer-dimer artifacts, and enzyme activity and fidelity. Because the dNTPs bind magnesium ions, the reaction mixture must contain an excess of $Mg^{2+}$ (0.5 ~ 2.5 mmol/L greater than the concentration of dNTPs).

### 6.2.5 DNA sequencing

DNA sequencing is mainly used for identification of new cDNA cloning, confirmation of cloning and mutation, inspection of new mutation, linkage, accuracy of PCR products, and analysis of non-coding gene sequences.

## 6.2.6 RNA analysis

One of the main uses of RNA isolation is for analysis of gene expression. In order to clarify the characteristics of gene regulation, it is necessary to learn the structures, quantity, level, size and synthesis rate etc. of RNA produced by gene transcription. Commonly used methods for analysis of RNA structure and quantity involve SI endonuclease analysis, ribonuclease protection experiments, primer extension, Northern blotting analysis. Nuclear run-off transcription analysis technique can be used to determine the number of activated RNA polymerase in a known gene in eukaryotic cells.

Experimental Manual in Medical Biochemistry

**Experiment 15**  Isolation of Eukaryotic Genomic DNA by Proteinase K-Phenol or NaI Method

## Protocol I  Proteinase K-Phenol Method

### Principle

Genomic DNA molecules exist as nucleoprotein in cells. The isolation principle of DNA is not only to separate DNA from complex with proteins, lipids and sugars, but also to maintain its integrity. This is the method of choice when high-molecular-weight DNA is required. Cells or tissues are lysed with SDS and proteinase K. EDTA chelates $Ca^{2+}$ and $Mg^{2+}$ to inhibit degradation of DNA by DNase. Phenol and chloroform/isoamyl alcohol extractions follow to remove most of the non-nucleic acid organic molecules (isoamyl alcohol gets rid of the foam). The DNA is precipitated out of solution with 70% ethanol, and resuspended in TE buffer (pH 8.0). Approximately 200 μg of mammalian DNA, 100 ~ 150 kb in length, is obtained from $5 \times 10^7$ cultured aneuploid mammalian cells. DNA obtained can be used for Southern blotting, PCR or for construction of genomic libraries.

### Reagents

1. Mammalian cell lysis buffer: 10 mmol/L Tris-Cl; 0.1 mol/L EDTA, pH 8.0; 0.5% SDS; 20 μg/ml RNase A (DNase free).
2. Proteinase K (20 mg/ml).
3. Phenol, equilibrated with 0.5 mol/L Tris-Cl (pH 8.0).
4. chloroform/isoamyl alcohol (24:1, V: V).
5. Sodium acetate (3 mol/L, pH 5.2).
   Dissolve 40.82 g NaAc ($3H_2O$) in 60 ml dd$H_2O$. Add enough glacial acetic acid to bring pH to 5.2. Bring final volume to 100 ml. Store at 4℃ after autoclave.
6. Anhydrous ethanol.
7. 70% ethanol.
8. 0.1 mol/L Tris-Cl (pH 8.0).
9. TE buffer (10 mmol/L Tris-Cl, 1 mmol/L EDTA, pH 8.0).
10. TBS (Tris-buffered saline, pH 7.4): Dissolve 8 g NaCl, 0.2g KCl and 3g Tris in 800 ml

# Chapter VI  Isolation, Purification and Identification of Nucleic Acids

ddH$_2$O, add HCl to pH 7.4, then add ddH$_2$O to 1 L.

## Procedure

1. According to the different samples, choose one of methods as step 1:
(1) Cell samples.
   ① Lyse Cells growing in monolayer cultures
   Take one batch of culture dishes, immediately remove the medium by aspiration. Wash the monolayers of cells twice with ice-cold TBS. Scrape the cells into 1 ml TBS, and then transfer the cell suspension to a centrifuge tube on ice. Recover the cells by centrifugation at 1500g for 10 min at 4℃. Resuspend the cells in 5~10 volumes of ice-cold TBS and repeat the centrifugation. Resuspend the cells in TE buffer at a concentration of $5 \times 10^7$ cells/ml. Add 1 ml lysis buffer per 0.1 ml cell suspension. Incubate the suspension for 1 h at 37℃, and then proceed immediately to step 2.
   ② Lyse cells growing in suspension cultures
   Transfer the cells to a centifuge tube and recover them by centrifugation at 1500g for 10 min at 4℃. Remove the supernatant. Wash the cells with ice-cold TBS and centrifugation, repeat twice. Remove the supernatant and gently suspend the cells in TE at a concentration of $5 \times 10^7$ cells/ml. Add 1 ml lysis buffer per 0.1 ml cell suspension. Incubate the solution for 1 h at 37℃ and then proceed immediately to step 2.
(2) Tissue samples
   Drop approx. 1 g of freshly excised tissue into liquid nitrogen in the stainless-steel container. Blend at top speed until the tissue is ground to a powder. Allow the liquid nitrogen to evaporate, and add the powdered tissue little by little to approx. 10 volumes (W/V) of lysis buffer in a beaker. Allow the powder to spread over the surface of the lysis buffer, and then shake the beaker to submerge the material. Transfer the suspension to a 50 ml centrifuge tube. Incubate the tube for 1 h at 37℃, and then proceed to Step 2.
(3) Blood samples
   ① To collect cells from freshly drawn blood
   Collect approx. 2 ml of fresh blood in tubes containing either ACD (Citric acid 0.48%, Sodium Citrate 1.32%, Dextrose 1.47%) or EDTA. Transfer the blood to a centrifuge tube and centrifuge at 1300g for 15 min at 4℃. Remove the supernatant. Use a Pasteur pipette to transfer the buffy coat carefully to a fresh tube and repeat the centrifugation. Discard the pellet of red cells. Remove residual supernatant and resuspend the buffy coat in 1.5 ml lysis buffer. Incubate the solution for 1 h at 37℃, and proceed to step 2.
   ② To collect cells from frozen blood samples
   Thaw the blood in a water bath at room temperature. Add an equal volume of phosphate-buffered saline and centrifuge the blood at 3500g for 15 min at room temperature. Remove

the supernatant, which contains lysed red cells. Resuspend the pellet in 1.5 ml of lysis buffer. Incubate the solution for 1 h at 37℃, and then proceed to step 2.

2. Add proteinase K (20 mg/ml) to a final concentration of 100 μg/ml. Use a glass rod to mix the solution gently into the viscous lysate of cells. Incubate the lysate in a water bath for 1 h at 37℃ followed by 3 h at 50℃. Swirl the viscous solution from time to time.

3. Cool the solution to room temperature and add an equal volume of phenol equilibrated with 0.1 mol/L Tris-Cl (pH 8.0). Gently mix the two phases by slowly turning the tube end-over-end for 10 min on a tube mixer or roller apparatus.

4. Separate the two phases by centrifugation at 5000g for 15 min at room temperature. Transfer the viscous aqueous phase to a fresh centrifuge tube. Repeat the extraction with phenol twice more if it is necessary.

5. Add an equal volume of chloroform/isoamyl alcohol (24:1), place on slowly rotating wheel for 15 min. Centrifuge 5000g for 10 min. Transfer the pooled aqueous phases to a fresh centrifuge tube and add 0.1 volume of 3 mol/L sodium acetate. Add 2.5 volumes of ice-cold anhydrous ethanol and swirl the tube until the solution is thoroughly mixed. The DNA immediately forms a precipitate.

6. Centrifuge 10000g for 15 min. Discard the supernatant. Wash the DNA precipitate twice with ice-cold 70% ethanol, and collect the DNA by centrifugation at 5000g for 5 min.

7. Remove as much of the 70% ethanol as possible. Store the pellet of DNA in an open tube at room temperature until the last visible traces of ethanol have evaporated. Add 0.1 ml of TE (pH 8.0) and place the tube on a rocking platform and gently rock the solution for 12~24 h at 4℃ until the DNA has completely dissolved. Store the DNA solution at 4℃.

8. Analyze the quality of the preparation of high-molecular-weight DNA by pulsed-field gel electrophoresis or by electrophoresis through a conventional 0.6% agarose gel. Purity of DNA product and the concentration of DNA could be detected by UV spectrophotometry.

## Notes

1. All glasswares and solutions (except organic solvents) should be sterilized. Gloves should be worn to avoid contamination of the experimental material and apparatus with skin exudates.

2. The pH of the phenol must be approx. 8.0 to prevent DNA from becoming trapped at the interface between the organic and aqueous phases.

3. When transferring the aqueous (upper) phase, it is essential to draw the DNA into the pipette very slowly to avoid disturbing the material at the interface and to minimize hydrodynamic shearing forces.

4. Do not allow the pellet of DNA to dry completely; desiccated DNA is very difficult to dissolve.

# Protocol II  Preparation of White Blood Cell DNA by NaI Method

## Principle

Nucleated blood cells are used to prepare genomic DNA. In the extraction solution, high concentration of NaI serves as detergent to dissolve nuclear membrane and separate DNA from histone. EDTA chelates $Ca^{2+}$ and $Mg^{2+}$ to inhibit degradation of DNA by DNase. Then chloroform/isoamyl alcohol precipitate proteins and dissolve lipids (isoamyl alcohol get rid of the foam). Isopropanol may be used to precipitate DNA in the water phase. Finally, DNA pellet is washed by 37% isopropanol or 70% ethanol, then dissolve in water or TE buffer (pH 8.0).

## Reagents

1. 1 × TE buffer (pH 8.0): (10 mmol/L Tris-Cl, 1 mmol/L EDTA-$Na_2$, sterilized).
2. 10% SDS.
3. 6 mol/L NaI (sodium Iodide).
4. Chloroform/isoamyl alcohol (24:1, V/V).
5. isopropanol.
6. 37% isopropanol.
7. 70% ethanol.
8. double distilled water (sterilized)(dd$H_2O$).

## Procedure

1. Pipette 0.1 ml peripheral blood to a 1.5 ml Eppendorf (Ep) tube, then centrifuge 10000 r/min for 1 min. Discard the upper aqueous layer and add 200 μl dd$H_2O$. Mix gently for 20 sec.
2. Add 200 μl NaI, Mix it by hand with a medium up and down motion for 20 sec.
3. Add 400 μl Chloroform/isoamyl alcohol. Mix it by hand thoroughly for 20 sec, and then centrifuge at 12 000 r/min for 12 min.
4. Remove 360 μl of upper aqueous layer carefully with an adjustable pipette and transfer it to a new 1.5 ml Ep tube. Add 200 μl isopropanol. Mix up by hand for 20 sec. After 15min at room temperature, centrifuge at 14 000 r/min at 4 ℃ for 12 min.
5. Discard the supernatant carefully and wash the pellet by adding 1 ml of 37 % isopropanol or 70% ethanol (don't vortex). Centrifuge 12 min at 14000 r/min, discard the supernatant carefully.

6. Remove the last trace liquid on the inner wall of tube by pipette or filter paper. Evaporate the isopropanol at room temperature for 15 min.
7. Add 50 μl TE buffer to dissolve the pellet gently. Employ agarose gel electrophoresis to determine the yield and purity of DNA isolation or store the samples at -20℃.

## Notes

1. Blood samples must be fresh. All performance should be done at 4℃.
2. The Ep tubes and tips must be sterilized.
3. Severely shaking will break DNA chain, so shaking should be gently either by hand or by vortexer.

Chapter VI  Isolation, Purification and Identification of Nucleic Acids

**Experiment title**
**Date**
**Observations and results**

**Discussion**

1. Genomic DNA molecules exist as _____ in cells.
2. On the process of isolating genomic DNA, the role of chloroform/ isoamyl alcohol is _____. The role of 37% Isopropanol or 70% ethanol is _____.
3. EDTA can inhibit degradation of DNA by DNase through _____.
4. In protocol II, high concentration of NaI serves as _____ and separate DNA from histone.
5. What is the isolation principle of DNA?

**Teacher's remarks**
**Signature**
**Date**

### Experiment 16　SDS-Alkaline Lysis of Plasmid DNA

**Principle**

This method exploits the difference in denaturation and renaturation characteristics of covalently closed circular plasmid DNA and chromosomal DNA fragments. Under alkaline conditions (at pH 11), both plasmid and chromosomal DNA are efficiently denatured. Rapid neutralization with a high-salt buffer such as potassium acetate in the presence of SDS has two effects that contribute to the overall effectiveness of the method. First, rapid neutralization causes the chromosomal DNA to base-pair in an intrastrand manner, forming an insoluble aggregate that precipitates out of solution. The covalently closed nature of the circular plasmid DNA promotes interstrand rehybridization, allowing the plasmid to remain in solution. Second, the potassium salt of SDS is insoluble, so the protein precipitates and aggregates, which assists in the entrapment of the high-molecular-weight chromosomal DNA. Separation of soluble and insoluble material is accomplished by a clearing method (e.g., filtration, magnetic clearing or centrifugation). The soluble plasmid DNA is ready to be further purified.

**Reagents**

1. Solution I : 25 mmol/L Tris-Cl, 10 mmol/L EDTA, 50 mmol/L glucose(pH 8.0).
   Add 1.0 mol/L Tris-Cl (pH 8.0) 1250 μl, 0.5 mol/L EDTA (pH 8.0) 1 ml, and Glucose 450 mg into 30 ml ddH$_2$O. Bring final volume to 50 ml. Store at 4℃ after autoclave.
2. Solution II : 0.2 mol/L NaOH, 1.0% SDS.
   Add 1.0 mol/L NaOH 20 ml, 20% SDS 5 ml into 50 ml ddH$_2$O. Bring final volume to 100 ml. Store at room temperature.
3. Solution III : 3 mol/L potassium acetate, pH 4.8 by acetic acid.
   Add 5 mol/L potassium acetate 60 ml into 10 ml ddH$_2$O. Add enough glacial acetic acid to bring pH to 4.8 (approx. 11 ml). Bring final volume to 100 ml.
4. 3 mol/L Sodium acetate, pH 5.2 by acetic acid.
5. Phenol, equilibrated with 0.5 mol/L Tris-Cl (pH 8.0).
6. Chloroform/ Isoamyl alcohol (24:1, $V:V$).
7. 70% Ethanol.
8. TE buffer (10 mmol/L Tris-Cl, 1 mmol/L EDTA, pH 8.0).

## Procedure

1. Centrifuge 1.5 ml of the culture at full speed in a microfuge tube to pellet the cells; discard the supernatant. (Select to repeat the procedure by adding an additional 1.5 ml from the same culture to the pelleted cells, and recentrifuge.)
2. Resuspend the cells in the 100 μl Solution I. Make sure that the cells are evenly suspended. (The solution should be cloudy with no obvious clumps).
3. Add 200 μl of freshly prepared solution II, mix gently (no vortex), and incubate 3~5 min on ice (solution should be clear). Do not incubate for longer than 5 min.
4. Add 150 μl of solution III, mix gently (no vortex), and incubate 5 min on ice.
5. Centrifuge 15 min at full speed in table-top centrifuge; pour off supernatant into 1.5 ml microfuge tube. Add RNase to a final concentration of 20 μg/ml and incubate the samples for 20 min at 37℃.
6. Extract with an equal volume of phenol equilibrated with 0.1 mol/L Tris-Cl (pH 8.0). Centrifuge for 5 min at full speed in microfuge (~12000 r/min). Transfer aqueous (upper) phase to new microfuge tube.
7. Extract the samples with an equal volume of chloroform/isoamyl alcohol (24:1). (Mix layers gently by hand to avoid shearing chromosomal DNA. Transfer aqueous (upper) phase to new microfuge tube.
8. Add 0.1 volume of 3 mol/L NaAc (pH 5.2) and add 2.5 volumes of ice-cold anhydrous ethanol, mix gently and incubate at -20℃ for 30 min to precipitate the DNA. Centrifuge 10 min at full speed in microfuge and pour off the supernatant.
9. Wash pellet with 1 ml ice-cold 70% ethanol; centrifuge 5 min in microfuge. Remove as much of the 70% ethanol as possible. Store the pellet of DNA in an open tube at room temperature until the last visible traces of ethanol have evaporated.
10. Dissolve pellet in 50 μl TE buffer.
11. When the procedure is complete, it is usually a good idea to heat the DNA sample to 68℃ for 10 min to inactivate any contaminating DNase. Then store the plasmid DNA at -20℃.

## Notes

1. Solution I may be prepared as a 10 × stock solution and stored -20℃ in small aliquots for later use. After the Solution I is added to the cells, make sure that the cells are thoroughly suspended by vortex.
2. Once the Solution II (SDS/NaOH mixture) has been added to the cells, it is imperative that the cells be treated gently; chromosomal DNA molecules are long and fragile, and vigorous treatment will readily damage the DNA molecules released from the cells.

# Experimental Manual in Medical Biochemistry

**Experiment title**
**Date**
**Observations and results**

**Discussion**

1. SDS-alkaline lysis method exploits the difference in _____ and characteristics of plasmid DNA and chromosomal DNA fragments.
2. Will plasmid and chromosomal DNA be denatured under alkaline conditions (such as pH 11)?
3. What is the role of high concentration of potassium acetate and SDS?

**Teacher's remarks**
**Signature**
**Date**

Chapter VI  Isolation, Purification and Identification of Nucleic Acids

## Experiment 17  Isolation of Total RNA by TRIzol Reagent

### Principle

TRIzol Reagent (U. S. Patent No. 5,346,994) is a ready-to-use reagent for the isolation of total RNA from cells and tissues, which includes a mono-phasic solution of phenol and guanidine isothiocyanate. During sample homogenization or lysis, TRIzol Reagent maintains the integrity of the RNA, while disrupting cells and dissolving cell components. Addition of chloroform followed by centrifugation separates the solution into an aqueous phase and an organic phase. RNA remains exclusively in the aqueous phase. After transfer of the aqueous phase, the RNA is recovered by precipitation with isopropyl alcohol. After removal of the aqueous phase, the DNA and proteins in the sample can be recovered by sequential precipitation. Precipitation with ethanol yields DNA from the interphase, and an additional precipitation with isopropyl alcohol yields proteins from the organic phase.

### Reagents

1. TRIzol Reagent.
2. Chloroform.
3. Isopropyl alcohol.
4. 75% Ethanol (in DEPC-treated water).
5. RNase-free water or 0.5% SDS solution [To prepare RNase-free water, draw water into RNase-free glass bottles. Add diethylpyrocarbonate (DEPC) to 0.01%. Let stand overnight and autoclave. The SDS solution must be prepared using DEPC-treated, autoclaved water.]

### Procedure

1. Homogenization
(1) Tissues

    Homogenize tissue samples in 1 ml TRIzol Reagent per 50 ~ 100 mg of tissue. The sample volume should not exceed 10% of the volume of TRIzol used for homogenization.

(2) Cells Grown in Monolayer

    Lyse cells directly in a culture dish by adding 1ml TRIzol Reagent to a 3.5 cm diameter dish, and passing the cell lysate several times through a pipette. The amount of

TRIzol Reagent added is based on the area of the culture dish (1 ml per 10 cm$^2$). An insufficient amount of TRIzol Reagent may result in contamination of the isolated RNA with DNA.

(3) Cells Grown in Suspension

Pellet cells by centrifugation. Lyse cells in TRIzol Reagent by repetitive pipetting. Use 1 ml of the reagent per $5 \sim 10 \times 10^6$ of animal, plant or yeast cells, or per $1 \times 10^7$ bacterial cells. Disruption of some yeast and bacterial cells may require the use of a homogenizer.

2. Phase separation

Incubate the homogenized samples for 5 min at 15 ~ 30℃ to permit the complete dissociation of nucleoprotein complexes. Add 0.2 ml of chloroform per 1 ml of TRIzol Reagent, cap sample tubes securely. Shake tubes vigorously by hand for 15 s and incubate them at 15 ~ 30℃ for 2 ~ 3 min. Centrifuge the samples at no more than 12000g for 15 min at 2 ~ 8℃. Following centrifugation, the mixture separates into a lower rid, phenol-chloroform phase, an interphase, and a colorless upper aqueous phase. RNA remains exclusively in the aqueous phase. The volume of the aqueous phase is about 60% of the volume of TRIzol Reagent used for homogenization.

3. RNA precipitation

Transfer the aqueous phase to a fresh tube, and save the organic phase if isolation of DNA or protein is desired. Precipitate the RNA from the aqueous phase by mixing with isopropyl alcohol. Use 0.5 ml of isopropyl alcohol per 1 ml of TRIzol Reagent used for the initial homogenization. Incubate samples at 15 ~ 30℃ for 10 min and centrifuge at no more than 12000g for 10 min at 2 ~ 8℃. The RNA precipitate, often invisible before centrifugation, forms a gel-like pellet on the side and bottom of the tube.

4. RNA wash

Remove the supernate. Wash the RNA pellet once with 75% ethanol, adding at least 1 ml of 75% ethanol per 1 ml of TRIzol Reagent used for the initial homogenization. Mix the sample by vortex and centrifuge at no more than 7500g for 5 min at 2 ~ 8℃.

5. Redissolving RNA

Dry the RNA pellet for 5 ~ 10 min in air. Do not dry the RNA by centrifugation under vacuum. It is important not to let the RNA pellet dry completely as this will greatly decrease its solubility. Dissolve RNA in RNase-free water or 0.5% SDS solution by passing the solution a few times through a pipette tip, and incubating for 10 min at 55 ~ 60℃. (Avoid SDS when RNA will be used in subsequent enzymatic reactions.)

## Notes

1. When working with TRIzol Reagent use gloves and eye protection (shield, safety goggles).

# Chapter VI  Isolation, Purification and Identification of Nucleic Acids

Avoid contact with skin or clothing. Use in a chemical fume hood. Avoid breathing vapor.

2. Isolation of RNA from small quantities of tissue ($1 \sim 10$ mg) or cells ($10^2 \sim 10^5$) samples: add 800 μl of TRIzol to the tissue or cells. Add 200 μg glycogen directly to the TRIzol (final concentration is 250 μg/ml). To reduce viscosity, shear the genomic DNA with 2 passes through a 26 guage needle prior to chloroform addition. The glycogen remains in the aqueous phase and is co-precipitated with the RNA. It does not inhibit first-strand synthesis at concentrations up to 4 mg/ml and does not inhibit PCR.

3. After homogenization and before addition of chloroform, samples can be stored at $-60$ to $-70$℃ for at least one month, the RNA precipitate (step 4, RNA wash) can be stored in 75% ethanol in $2 \sim 8$℃ for at least one week, or at least one year at $-5$ to $-20$℃.

**Experiment title**
**Date**
**Observations and results**




**Discussion**

1. TRIzol Reagent includes a mono-phasic solution of _____ and _____.
2. RNA is high sensitive to RNA enzyme (RNase), which has very stable biological activity. RNase activity does not require _____, and not influenced by _____.
3. How to control the contaminating RNase for RNA isolation?

**Teacher's remarks**
**Signature**
**Date**

# Chapter VI  Isolation, Purification and Identification of Nucleic Acids

## Experiment 18   Identification of DNA by UV-Spectrophotometry and Gel Electrophoresis

## Part I   Identification of DNA by UV-Spectrophotometry

### Principle

With the use of UV spectrophotometry, the quantitative analysis of nucleic acids has established itself as a routine method. Aqueous buffers with low ion concentrations (e. g. TE buffer) are ideal for this method. The DNA concentration is determined by measuring absorption at 260 nm ($A_{260}$) against blank and then calculating the concentration of the measured sample using according factors. The absorption of 1 OD ($A$) is equivalent to approximately 50 μg/ml dsDNA or approximately 30 μg/ml for oligonucleotides. Purity determination of DNA Interference by contaminants can be recognized by the calculation of "ratio". In the case of pure samples, the ratio $A_{260}/A_{280}$ should be approximately 1.8, and the ratio $A_{260}/A_{230}$ should be approximately 2.2. Phenol has an absorbance maximum of 270 nm but the absorbance spectrum overlaps considerably with that of nucleic acids. If there is phenol contamination in your DNA sample, the absorbance at 260 nm will be high, giving a false measure of DNA concentration.

You can calculate the concentration of the DNA in your sample as follows:

DNA concentration (μg/ml) = ($A_{260}$) × (dilution factor) × (50 μg DNA/ml) / (1 $A_{260}$ unit)

### Reagents

1. Distilled $H_2O$ ($dH_2O$).
2. DNA solution (unknown concentration).

### Procedure

1. Turn on machine using the switch in the back of the machine and wait for it to warm up. Open the lid on the top of the machine. Place the blank cuvette in the cuvette chamber in front of the light source, then set wavelength to '260'. Wash quartz cuvettes thoroughly before use and if two cuvettes are used then make sure they are a 'matching' pair.

2. Insert a cuvette within 1 ml of water into machine and press 'autozero'.
3. Put 5 μl of DNA with unknown concentration into a clean matching cuvette, add 1 ml of water and mix by covering with parafilm and inverting.
4. Insert cuvette into machine and note reading i. e. 0.30.
5. Follow steps 3 again but this time set wavelength to 280, note reading.

Take the $A_{260}$ readings and multiply the figure by 10 to give the concentration of DNA per ml. i. e. $0.30 \times 10 = 3.0$ mg/ml.

Take $A_{280}$ readings, divide $A_{260}$ by the $A_{280}$, and this should give a figure around 1.8. If it is less then 1.7, the DNA is not pure and sample should be purified using phenol/chloroform.

## Part II  Identification of DNA by Agarose Electrophoresis of DNA

### Principle

Agarose gel electrophoresis is employed to check the progression of a restriction enzyme digestion, to quickly determine the yield and purity of a DNA isolation or PCR reaction, and to size fractionate DNA molecules, which then could be eluted from the gel. Prior to gel casting, dried agarose is dissolved in buffer by heating and the warm gel solution then is poured into a mold, which is fitted with a well-forming comb. The percentage of agarose in the gel varied and gels should be prepared depending on the expected size(s) of DNA fragment(s).

Agarose gels are often run in a horizontal configuration. Electrophoresis usually is at 150~200 mA for 0.5~1 h at room temperature, depending on the desired separation. Size markers are co-electrophoresed with DNA samples, when appropriate for fragment size determination.

Nucleic acids are typically visualized by staining with ethidium bromide or SYBR gold. Ethidium bromide is included in the gel matrix to enable fluorescent visualization of the DNA fragments under UV light.

### Reagents

1. 5 × TBE buffer (89 mmol/L Tris base, 89 mmol/L boric acid, 2 mmol/L EDTA, pH 8.3).
2. 6 × DNA loading buffer (30% Glycerol, 0.25% Bromophenol blue, 0.25% green xylene).
3. 5 mg/ml Ethidium bromide (EB).
4. 1.2% agarose gel: Add agarose 1.2 g and 0.5 × TBE 100 ml into 0.5 L flask, and heat them in a microwave for 2~4 min until the agarose is dissolved.

5. DNA samples.
6. DNA size standard (DNA Marker).

## Procedure

1. Assemble the gel casting tray and comb. The comb should not touch the bottom of the tray.
2. Melt the agarose gels using a microwave. When the agarose solution has cooled to about 50~60℃, pour solution into the casting tray, ensuring that no bubbles get into the gel.
3. Allow the gel to cool. It will solidify and become slightly opaque within 20~30 min.
4. Adding 0.5 × TBE buffer to cover the gel by about a half of centimeter.
5. Carefully remove the comb by lifting it gently at one end, tilting the comb as it comes out. Pulling the comb straight up creates a vacuum in the wells, which tends to lift the whole gel out of the tray. Ensure that the wells are submerged and filled with buffer.
6. Prepare the DNA samples for loading using DNA loading buffer.
7. Pipette 10 μl of each sample into individual wells with a gel loading tip.
8. Once all the samples are loaded, place the cover on the gel apparatus. Connect the leads so that the red (positive) lead is at the end of the gel to which the DNA will migrate and the black (negative) lead is at the end of the gel containing the wells. Turn on the power supply and set according to the instructor's guidelines. Check the gel after a few min. If the tracker migrates in the wrong direction, turn off the power supply and switch the leads. Run at a constant voltage of 120 V. Caution: Do not remove the lid from the gel apparatus without disconnecting the power leads.
9. Turn off the power supply and disconnect the power leads when the blue tracking dye has migrated about 75% of the distance to the end of the gel (usually within 60~90 min). Caution, the gel is very slippery!!
10. Carefully transfer the gel into a plastic dish and add enough ethidium bromide (EB) staining solution to cover the gel. Set in a dark drawer for 30 min.
11. Visualize the DNA fragments on a long wave UV light box.

## Notes

1. When preparing agarose for electrophoresis, it is best to sprinkle the agarose into room-temperature buffer, swirl, and let sit at least 1 min before microwaving. This allows the agarose to hydrate first, which minimizes foaming during heating.
2. The minimum amount of DNA detectable by EB on a 3-mm-thick gel and a 5-mm-wide lane is 1 ng. Loading DNA in the smallest volume possible will result in sharper bands.
3. Migration of DNA is retarded and band distortion can occur when too much buffer covers the gel. The slower migration results from a reduced voltage gradient across the gel.
4. You can preserve DNA in agarose gels for long-term storage using 70% ethanol.

# Part III  Polyacrylamide Gel Electrophoresis of DNA

## Principle

Non-denaturing polyacrylamide gel is used for the separation and purification of small double-stranded DNA fragments ( <1000 bp). The mobility of most double-stranded DNA is roughly inversely proportional to the logarithm of the molecular weight, but it is also affected by the base composition and sequence. The migration of DNA molecules with same size may be different due to the difference in spatial structure. Therefore, non-denaturing polyacrylamide gel can not be used to determine the size of double-stranded DNA.

## Reagents

1. Acrylamide: bisacrylamide stock solution (29:1).
2. 10% Ammonium persulfate.
3. 6 × Gel-loading buffer.
4. 5 × TBE electrophoresis buffer.
5. 10% TEMED.

## Procedure

1. Wash the glass plates and spacers in warm detergent solution and rinse them well, first in tap water and then in deionized $H_2O$. Hold the plates by the edges or wear gloves, so that oils from the hands do not become deposited on the working surfaces of the plates. Rinse the plates with ethanol and set them aside to dry.
2. Assemble the glass plates with spacers. Taking into account the size of the glass plates and the thickness of the spacers, calculate the volume of gel required. Prepare the gel solution with the desired polyacrylamide percentage according to the table below, which gives the amount of each component required to make 10 ml.

Table 6-2  Recipes for polyacrylamide gels of indicated concentrations in 1 × TBE

| Gel concentration (%) <br> Reagents (ml) | 3.5 | 5.0 | 8.0 | 12.0 | 20.0 |
| --- | --- | --- | --- | --- | --- |
| 30% Acrylamide stock solution | 1.16 | 1.66 | 2.66 | 4.0 | 6.66 |
| $H_2O$ | 6.7 | 6.2 | 5.2 | 3.86 | 1.2 |
| 5 × TBE | 2.0 | 2.0 | 2.0 | 2.0 | 2.0 |
| 10% Ammonium persulfate | 0.1 | 0.1 | 0.1 | 0.1 | 0.1 |

3. Add 50 μl of 10% TEMED for each 10 ml of acrylamide: bis solution, mix the solution by gentle swirling, and then pour the gel solution into the space. (Wear gloves. Work quickly to complete the gel before the acrylamide polymerizes.)
4. Immediately insert the appropriate comb into the gel, being careful not to allow air bubbles to become trapped under the teeth. Make sure that no acrylamide solution is leaking from the gel mold. Allow the acrylamide to polymerize for 30~60 min at room temperature.
5. When ready to proceed with electrophoresis, squirt 1 × TBE buffer around and on top of the comb and carefully pull the comb from the polymerized gel. Use a syringe to rinse out the wells with 1 × TBE.
6. Fill the reservoirs of the electrophoresis tank with electrophoresis buffer. Use a Pasteur pipette or syringe needle to remove any air bubbles trapped beneath the bottom of the gel.
7. Use a syringe to flush out the wells once more with 1 × TBE. Mix the DNA samples with the appropriate amount of 6 × gel-loading buffer. Load the mixture into the wells.
8. Connect the electrodes to a power pack (positive electrode connected to the bottom reservoir), turn on the power, and begin the electrophoresis run.
9. Run the gel until the marker dyes have migrated the desired distance. Turn off the electric power, disconnect the leads, and discard the electrophoresis buffer from the reservoirs.
10. Detach the glass plates. Lay the glass plates on the bench. Use a spacer or plastic wedge to lift a corner of the upper glass plate. Check that the gel remains attached to the lower plate. Pull the upper plate smoothly away. Remove the spacers.
11. Carefully transfer the gel into a plastic dish and add EB staining solution to cover the gel. Set in a dark drawer for 30 min. Visualize the DNA fragments by a long wave UV light.

## Notes

1. Polyacrylamide gels are poured and run in 0.5 × or 1 × TBE at low voltage (1~8 V/cm) to prevent denaturation of small fragments of DNA. Other electrophoresis buffers such as 1 × TAE can be used, the gel must be run more slowly in 1 × TAE, which does not provide as much buffering capacity as TBE. For electrophoresis runs greater than 8 h, we recommend that 1 × TBE buffer be used because adequate buffering capacity is available.
2. The glass plates must be free of grease spots to prevent air bubbles from forming in the gel.
3. It is important to use the same batch of electrophoresis buffer in both of the reservoirs and in the gel. Small differences in ionic strength or pH produce buffer fronts that can greatly distort the migration of DNA.
4. Usually, approx. 20~100 μl of DNA sample is loaded per well depending on the size of the slot. Do not attempt to expel all of samples from the loading device, as this almost always produces air bubbles that blow the sample out of the well. In many cases, the same device can be used to load many samples, provided it is thoroughly washed between each loading.

However, it is important not to take too long to complete loading the gel; otherwise, the samples will diffuse from the wells.

5. Non-denaturing polyacrylamide gels are usually run at voltages between 1 V/cm and 8 V/cm. If electrophoresis is carried out at a higher voltage, differential heating in the center of the gel may cause bowing of the DNA bands or even melting of the strands of small DNA fragments. Therefore, with higher voltages, gel boxes that contain a metal plate or extended buffer chamber should be used to distribute the heat evenly. Alternatively, use a gel-temperature-monitoring strip.

## Part IV  Gel Extraction of DNA

### Principle

Crushing gel, immersing in eluate, and centrifugation is one of methods for recovering DNA from gels. Obtained DNA usually does not contain enzyme inhibitors and other impurities poisonous for transfection or microinjection. The recovery rate is 30% to 90% depending on the size of DNA fragment. This method is widely used for isolating single, double-stranded DNA and oligonucleotides from gels, which are suitable as hybridization probe, templates for chemical and enzymatic sequencing. Fragments from 200 bp to 10 kb the agarose purification is ideal. For smaller fragments (20~400 bp) the acrylamide purification is preferred.

### Reagents

1. Crush and Soak Solution (500 mmol/L $NH_4Ac$, 0.1% SDS, 0.1 mmol/L EDTA).
   Add 3.3 g $NH_4Ac$, 0.1 g SDS, 20 ml 500 mmol/L EDTA into $ddH_2O$ and bring up to 100 ml, store at room temperature.
2. 3 mol/L NaAc, pH 5.2.
   24.6 g anhydrous sodium acetate, adjust pH to 5.2 with acetic acid and bring up to 100 ml with $ddH_2O$, store at room temperature.
3. DMCS treated glass wool (Alltech Assoc. Inc. #4037, 50 g).
4. Blue tips with melted tips to serve as pestle for crushing acrylamide.

### Procedure

#### I. Agarose gels

1. Prepare spin columns by cutting off the cap of a 0.5 ml eppendorf (Ep) tube and forming a hole in the bottom with a hot 18 Ga needle. Fill this "mini-column" with a small ball of DMCS treated glass wool and pack down with a pipet tip.

Chapter VI  Isolation, Purification and Identification of Nucleic Acids

2. Cut out the desired band from an agarose gel and place in a spin column inside a 1.5 ml Ep tube with the top cut off.
3. Spin at 6000 r/min for 10 min.
4. Phenol/chloroform extracts the flow through and Ethanol precipitates with glycogen or tRNA and 10% V/V of 3 mol/L NaAc.
5. Wash and dry, resuspend in 20 μl TE.

## II. PAGE

1. Run a 4% ~ 6% PAGE in 1 × TBE, stain in EB (1 ~ 10 mg/ml) and cut out the desired band. Crush the acrylamide with a p1000 tip with a melted end to resemble a pestle for the eppendorf "mortar."
2. Add 1 ml crush and soak solution and incubate overnight at 37℃.
3. Spin in the microfuge for 10 min at 14,000 r/min. Remove as much liquid as possible and add another 0.5 ml of crush and soak solution.
4. Repeat the spin and pool the recovered supernatant.
5. Add 0.1 volume of 3 mol/L NaAc, 2.5 volumes of Ethanol and carrier (see above).
6. Spin as usual, wash and dry. Resuspend in 20 μl TE.

**Experiment title**
**Date**
**Observations and results**

**Discussion**

1. The absorption of 1 OD ($A$) is equivalent to approximately _____ dsDNA.
2. The ratio $A_{260}/A_{280}$ is used to estimate the purity of nucleic acid. Pure DNA should have a ratio of approximately ____, whereas pure RNA should give a value of approximately ____.
3. Agarose gels are often run in a _____ configuration, while polyacrylamide gels are often run in a _____ configuration.
4. Non-denaturing polyacrylamide gel is used for the separation and purification of _____ ____.

**Teacher's remarks**
**Signature**
**Date**

Chapter VI Isolation, Purification and Identification of Nucleic Acids

## Experiment 19  Identification of RNA by UV-Spectrophotometry and Formaldehyde Denaturating Agarose Gel Electrophoresis

### Part I  Identification of RNA by UV-Spectrophotometry

### Principle

The RNA concentration is determined by measuring absorption at 260 nm ($A_{260}$) against blank and then calculating the concentration of the measured sample using according factors. The absorption of 1 OD (A) is equivalent to approximately 40 μg/ml for RNA. In the case of pure samples, the ratio $A_{260}/A_{280}$ should be approximately 2.0.

You can calculate the concentration of the RNA in your sample as follows:

RNA concentration (μg/ml) = $A_{260}$ × (dilution factor) × 40/ (1 $A_{260}$ unit)

### Reagents

1. Distilled $H_2O$ (d$H_2O$).
2. RNA solution (unknown concentration).

### Procedure

1. See experiment 18. The only difference is the dilution step (step 3): put 4 μl of RNA with unknown concentration into 1 ml of water and mix.
2. Take $A_{280}$ readings, divide $A_{260}$ by the $A_{280}$, and this should give a figure around 2.0. If it is less then 1.9, the RNA is not pure and sample should be purified using phenol/chloroform.

### Part II  Identification of RNA by Formaldehyde Denaturating Agarose Gel Electrophoresis

### Principle

A denaturing gel system is suggested because most RNA forms extensive secondary

153

structure via intramolecular base pairing, and this prevents it from migrating strictly according to its size. The mobility of RNA (or DNA) will not be affected by their base composition and conformation in the presence of base pairing inhibitors (formaldehyde, urea or formamide), and the relationship between mobility and common logarithm of molecular weight will be inversely proportional. Denaturing agarose gel electrophoresis can be used for RNA fractionation, Northern blot, measuring the length of RNA (or DNA), and the recovery of oligonucleotide.

## Reagents

1. Formaldehyde.
2. Formamide (Purchase or prepare a distilled-deionized preparation of this reagent and store in small aliquots under nitrogen at-20℃).
3. 5 × MOPS buffer (0.1 mol/L MOPS, pH 7.0, 40 mmol/L NaAc, 5 mmol/L EDTA). Dissolve 20.6 g MOPS in 800 ml DEPC-treated 50 mmol/L NaAc solution, add enough 2 mol/L NaOH to adjust pH to 7.0, and then add 10 ml 0.5 mol/L EDTA (pH 8.0). Bring final volume to 1 L by adding $ddH_2O$. Store at 4℃ in the dark after autoclave.
4. 10 × formaldehyde gel-loading buffer (50% glycerol, 1 mmol/L EDTA pH 8.0, 0.25% bromphenol blue, 0.25% green xylene), treated with 0.1% DEPC water and autoclave.
5. Ethidium bromide (EB, 0.5 μg/ml).

## Procedure

1. Set up the denaturation reaction. In sterile microfuge tubes mix:
   | | |
   |---|---|
   | RNA (up to 20 μg) | 4.0 μl |
   | 5 × MOPS buffer | 2.0 μl |
   | formaldehyde | 4.0 μl |
   | formamide | 10.0 μl |
2. Incubate the RNA solutions for 15 min at 65℃. Chill the samples for 5 ~ 10 min on ice, and then centrifuge them for 5 sec to deposit all of the fluid in the bottom of the tubes. Add 2 μl of 10 × formaldehyde gel-loading buffer to each sample and place the tubes on ice.
3. Install the agarose/formaldehyde gel in a horizontal electrophoresis box. Add sufficient 1 × MOPS buffer to cover the gel to a depth of approx. 1 mm. Run the gel for 5 min at 5 V/cm, and then load the RNA samples into the wells of the gel.
4. Run the gel submerged in 1 × MOPS buffer at 4 ~ 5 V/cm until the bromophenol blue has migrated at least 2 ~ 3 cm into the gel, or as far as 2/3 the length of the gel (1 ~ 2 h).
5. Immerse the gel in 0.5 μg/ml of EB solution for 10 min after electrophoresis, and then destain in 10 mol/L $MgCl_2$ for 5 min. Visualize the RNAs by placing the gel on a UV transilluminator.

# Chapter VI  Isolation, Purification and Identification of Nucleic Acids

## Notes

1. Formaldehyde is toxic through skin contact and inhalation of vapors. Manipulations involving formaldehyde should be done in a chemical fume hood.
2. Intact total RNA run on a denaturing gel will have sharp 28S and 18S rRNA bands (eukaryotic samples). The 28S rRNA band should be approximately twice as intense as the 18S rRNA band. This 2:1 ratio (28S:18S) is a good indication that the RNA is intact. Partially degraded RNA will have a smeared appearance, will lack the sharp rRNA bands, or will not exhibit a 2:1 ratio.

Experimental Manual in Medical Biochemistry

**Experiment title**
**Date**
**Observations and results**

**Discussion**

1. Why is the denaturing gel system suggested for RNA electrophoresis?
2. Please give some base pairing inhibitors.
3. After total RNA isolation, it is required to identify the integrity of RNA by denaturing gel electrophoresis. Can we see mRNA bands under UV light by staining with ethidium bromide?

**Teacher's remarks**
**Signature**
**Date**

# Chapter VI  Isolation, Purification and Identification of Nucleic Acids

## Experiment 20  Polymerase Chain Reaction (PCR) and Reverse Transcription-PCR

## Protocol I  Polymerase Chain Reaction (PCR)

### Principle

The polymerase chain reaction (PCR) is a rapid procedure for in vitro enzymatic amplification of a specific segment of DNA. The purpose of a PCR is to make a huge number of copies of a gene. PCR is based on the DNA polymerization reaction. A primer and dNTPs are added along with a DNA template and the DNA Taq polymerase. In addition to using a primer that sits on the 5' end of the gene and makes a new strand in that direction, a primer is made to the opposite strand to go in the other direction. The amplification reaction involves three steps: ① denaturation: template DNA is heated to denature the double-stranded DNA molecules, making them single-stranded; ② annealing: the reaction mix is then cooled, allowing primers to anneal to complementary sequences on opposite strands of the template DNA (by hydrogen bonding between complementary bases: A-T, G-C) flanking the DNA segment to be amplified; ③ extension: the reaction is then brought to an intermediate temperature, and, using free deoxyribonucleotides added to the reaction mixture, DNA polymerase extends these primers from their 3' ends toward each other. This three-step process is repeated for 20 ~ 40 cycles, resulting in the production of many copies of the template DNA.

### Reagents

1. Diluted DNA template.
2. dNTP mix solution (2 mmol/L each dATP, dTTP, dCTP, dGTP) (pH 7.5).
3. 10 × PCR Buffer mix (100 mmol/L Tris-Cl, pH 8.3, 500 mmol/L KCl, 1.5 mmol/L $MgCl_2$).
4. Sense Primer (50 μmol/L).
5. Antisense Primer (50 μmol/L).
6. Taq DNA polymerase (2 ~ 5 U/μl).
7. Nuclease-free water.

Experimental Manual in Medical Biochemistry

## Procedure

1. Primers design: The selection of primers is very important to the efficiency of the reaction. Usually the primers are custom synthesized based on the sequence of the DNA that is being amplified. Two primers would have to be made for each of the inserts.
2. Take three 0.2 ml thin-walled microcentrifuge tubes and mark them with blank, positive and sample. Add the following reagents into each of the tubes: (you could add the Master mix buffer and Taq DNA polymerase, The Master Mix contains all of the components necessary to make new strands of DNA in the PCR process. It could be obtained from a variety of commercial companies).

   | | |
   |---|---|
   | 36.0 μl | $H_2O$ |
   | 5.0 μl | 10 × Buffer mix |
   | 1.0 μl | each primers (50 μM) |
   | 5.0 μl | dNTPs |
   | 1.0 μl | Taq DNA polymerase (2.5 U) |
   | 1.0 μl | diluted DNA template (0.05 ~ 1 μg) only put into sample tube |
   | 50.0 μl Total volume | |

3. Place these three tubes in the thermocycler and start to run the PCR program for the amplification.

   | | | | |
   |---|---|---|---|
   | 94 ℃ | 5 min | | DNA initial denaturation |
   | 25 cycles | 94 ℃ | 30 s | denaturation |
   | | 58 ℃ | 40 s | annealing |
   | | 72 ℃ | 1 min | extension |
   | 72 ℃ | 5 min (for final extension) | | |
   | 4 ℃ | forever (for storage) | | |

4. Analyze a 10 μl aliquot of PCR products by running on a 1% ~ 2% agarose gel electrophoresis.

## Notes

1. The essential criteria for any DNA sample are that it contains at least one intact DNA strand encompassing the region to be amplified and that any impurities are sufficiently diluted so as not to inhibit the polymerization step of the PCR reaction. Usually a 1:5 dilution of the sample with water is sufficient to dilute out any impurities.
2. PCR involves preparation of the sample, the primers and PCR mixture system, followed by detection and analysis of the reaction products.
3. After the first few cycles, most of the product DNA strands made are the same length as the distance between the primers. After PCR, aliquots of the mixture typically are loaded onto

an agarose gel and electrophoresed to detect amplified product, note the extent to which you diluted the DNA in the reactions. Reaction only has about 1 picogram of DNA template. For this reason, on the agarose gel you should only see the amplified fragment.

## Protocol II  Reverse Transcription-PCR

### Principle

Reverse transcription-PCR (RT-PCR) is the most sensitive method for mRNA detection and quantitation currently available. Compared to the two other commonly used techniques for quantifying mRNA levels, Northern blot analysis and RNase protection assay, RT-PCR can be used to quantify mRNA levels from much smaller samples. Traditionally RT-PCR involves two steps: the RT reaction and PCR amplification. RNA is first reverse transcribed into cDNA using a reverse transcriptase, and the resulting cDNA is used as templates for subsequent PCR amplification using primers specific for one or more genes. RT-PCR can also be carried out as one-step RT-PCR in which all reaction components are mixed in one tube prior to starting the reactions. Although one-step RT-PCR offers simplicity and convenience and minimizes the possibility for contamination, the resulting cDNA cannot be repeated used as in two step RT-PCR. Reverse Transcription here is carried out with the SuperScript First-Strand Synthesis System for RT-PCR.

### Reagents

1. Oligo(dT)$_{12-18}$ (100 μg/ml) in TE (pH 8.0) or random primers.
2. dNTP mix solution (10 mmol/L) containing all four dNTPs (pH 7.5).
3. SuperScript II reverse transcriptase.
4. 5 × First Strand Buffer (250 mmol/L Tris-Cl pH 8.3, 375 mmol/L KCl, 15 mmol/L MgCl$_2$). This buffer is supplied with the SuperScript Reverse Transcriptase.
5. 0.1 mol/L DTT.
6. 10 × PCR Buffer (200 mmol/L Tris-HCl pH 8.4, 500 mmol/L KCl).
7. 50 mmol/L MgCl$_2$.
8. Sense Primer (10 μmol/L) for amplification of cDNA by PCR.
9. Antisense Primer (10 μmol/L) for amplification of cDNA by PCR.
10. Taq DNA polymerase (2~5 U/μl).
11. Nuclease-free water.

Experimental Manual in Medical Biochemistry

## Procedure

1. First Strand cDNA Synthesis: A 20 μl reaction volume can be used for 1~5 μg of total RNA or 50~500 ng of mRNA. Add the following components to a nuclease-free microfuge tube:

   1 μl Oligo(dT)$_{12-18}$ or 50~250 ng of random primers
   1~5 μg Total RNA
   Nuclease-free water to 12 μl

   Heat mixture to 70℃ for 10 min and quick chill on ice. Collect the contents of the tube by brief centrifugation and add:

   4 μl 5 × First Strand Buffer.
   2 μl 0.1 mol/L DTT
   1 μl 10 mmol/L dNTP Mix

2. Mix contents of the tube gently and incubate at 42℃ for 2 min.

3. Add 1 μl (200 units) of SuperScript II, mix by pipetting gently up and down. Incubate 50 min at 42℃.

4. Inactivate the reaction by heating at 70℃ for 15 min, and then chill on ice. Store the 1st strand cDNA at -20℃ until use for PCR.

5. Add the following to a PCR reaction tube for a final reaction volume of 100 μl:

   10 μl 10 × PCR Buffer
   3 μl 50 mmol/L MgCl$_2$
   2 μl 10 mmol/L dNTP Mix
   2 μl sense Primer (10 μmol/L)
   2 μl antisense Primer (10 μmol/L)
   1 μl Taq DNA polymerase (2~5 U/μl)
   2 μl cDNA (from first strand reaction, preferably RNase H-treated)
   80 μl Autoclaved, distilled water

6. Mix gently. Heat reaction to 94℃ for 3 min to denature.

7. Perform 15~40 cycles of PCR.

   Denaturation: 45 sec at 94℃
   Annealing: 45 sec at 55℃
   Extension: 1 min at 72℃
   Final extension 10 min at 72℃

   Times and temperatures may need to be adapted to suit the particular reaction conditions.

8. Analyze 5 μl of the PCR product by agarose gel electrophoresis. The rest is stored at -20℃.

Chapter VI  Isolation, Purification and Identification of Nucleic Acids

**Notes**

1. The cDNA from first strand reaction can be used as a template for amplification in PCR. However, amplification of some PCR targets (those >1 kb) may require the removal of RNA complementary to the cDNA. To remove RNA complementary to the cDNA, add 1 μl (2 units) of E. coli RNase H and incubate 37℃ for 20 min.
2. Taq DNA polymerase is the standard and appropriate enzyme for the amplification stage of most forms of RT-PCR. However, where elongation from 3′-mismatched primers is suspected, a thermostable DNA polymerase with 3′-5′ proofreading activity may be preferred.
3. Use only 10% of the first strand reaction for PCR. Adding larger amounts of the first strand reaction may not increase amplification and may result in decreased amounts of PCR product.

# Experimental Manual in Medical Biochemistry

**Experiment title**
**Date**
**Observations and results**

## Discussion

1. There are three major steps in a PCR, which include _____, _____, and _____.
2. The purpose of a PCR is _____.
3. Traditionally RT-PCR involves two steps: _____ and _____.
4. What is the principle of PCR?

**Teacher's remarks**
**Signature**
**Date**

# Chapter VII
# Genetic Engineering

The term genetic engineering means the deliberate modification of an organism's genetic information by directly changing its nucleic acid genome and it is accomplished by a collection of methods known as recombinant DNA technology. It uses the techniques of molecular cloning to alter the structure and characteristics of genes directly. There are a number of ways through which genetic engineering is accomplished. Essentially, the process has four main steps: isolation of the genes of interest, insertion of the genes into a transfer vector, transformation of cells of organism to be modified, separation of the genetically modified organism from those that have not been successfully modified. Genetic engineering makes genetic information transfer freely among different species, even between prokaryotic and eukaryotic, plant and animals, etc. The promise of its application for medicine, agriculture, and industry is great, yet the potential risks of this technology are not completely known and may be considerable.

## 7.1 The Process of Gene Cloning

Cloning is the process of making an identical copy of something. In biology, it collectively refers to processes used to create copies of DNA fragments (gene cloning), cells (cell cloning), or organisms. The term also covers when organisms such as bacteria, insects or plants reproduce asexually. Although polymerase chain reaction technique can clone large quantities of DNA fragments *in vitro*, in biology 'gene cloning' more often refers to the procedure of isolating a defined DNA sequence and obtaining multiple copies of it *in vivo*. It is a process by which large quantities of a specific, desired gene or section of DNA may be copied once the desired DNA has been isolated. It is often used to get enough amount of gene for further analysis. Although gene cloning is frequently employed to amplify DNA fragments containing special target genes, it can also be used to amplify any DNA sequence such as promoters, non-coding sequences and randomly fragmented DNA. It is utilized in a wide array of biological experiments and practical applications such as large scale protein production.

### 7.1.1 Fragmentation/Isolation

Initially, the gene of interest which to be inserted into the organism needs to be identified and isolated to provide a DNA segment of suitable size, usually by using existing knowledge of the various functions of genes. DNA information can be obtained from cDNA or genome DNA libraries. One can extract the genome DNA or total RNA from an organism, then isolate the fragment of interest by cleaving the DNA or cDNAs into fragments. Alternatively, one also can directly synthesize the desired DNA fragment by assembly PCR and DNA synthesizing techniques. If necessary, i. e. for insertion of eukaryotic genomic DNA into prokaryotes, further modification may be carried out such as removal of introns or ligating with prokaryotic promoters. The gene or DNA that is desired then is treated with restriction enzymes for further cloning with suitable vectors.

### 7.1.2 Ligation/Insertion

Subsequently, a ligation procedure is used once the gene of interest is isolated. Joining linear DNA fragments together with covalent bonds is called ligation. The vector which is frequently circular needs to be linearized before the ligation. Then it is treated with the same restriction endonuclease as the gene of interest to produce identical sticky ends or be supplemented to special linkers at the ends. After that, both materials are incubated together under appropriate conditions with an enzyme called DNA ligase, then, the recombinant DNA is formed.

### 7.1.3 Transformation/Transfection

Following ligation, the vector with the insert of target gene is transformed into host cells to replicate. A number of alternative transformation techniques are available, such as chemical sensitivation of cells, microinjection, electroporation, bacterial transformation or using viruses as vectors etc. Depending on the vector used, it can be complex or simple.

### 7.1.4 Screening/Selection

Finally, the transformed cells are cultured and screened. As the aforementioned procedures are of particularly low efficiency, there is a need to identify the cells that have been successfully transformed with the vector construct containing the desired insertion sequence in the required orientation. Modern cloning vectors usually include selectable antibiotic resistance markers, which allow only cells in which the vector has been transformed, to grow on agar dishes with the proper antibiotics. Additionally, the cloning vectors may contain color selection markers which provide blue/white screening ($\alpha$-factor complementation) on X-gal medium. Nevertheless, these selection steps do not absolutely guarantee that the DNA insert is correctly present in the cells obtained. Further investigation of the resulting colonies is required to confirm whether the cloning is successful. This may be accomplished by means of PCR, restriction fragment analysis and/or DNA sequencing, etc., such as testing with DNA probes that can stick to the gene of interest that was supposed to have been transplanted.

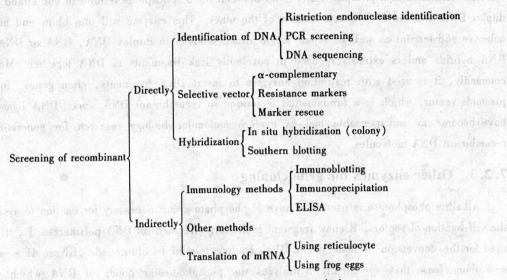

Fig. 7-1 Recombinant screening methods

## 7.2 Enzymes for Gene Cloning

### 7.2.1 Restriction endonucleases

There are three general types of restriction endonuclease. Type I and III cleave DNA away from recognition sites. Type II restriction endonucleases cleave DNA at specific recognition sites. The type II enzymes can be used to prepare DNA fragments containing specific genes or portions of genes in gene cloning techniques. For example, the restriction endonuclease *Eco*R I, isolated by Herbert Boyer in 1969 from *E. coli*, cleaves the DNA between G and A in the base sequence GAATTC. The main advantage of restriction enzymes is their ability to cut DNA strands reproducibly in the same places. This property is the basis of many techniques used to analyze genes and their expressions. Moreover, many restriction enzymes make staggered cuts in the two DNA strands, leaving single-stranded overhangs, or sticky ends, that can form base-pair together briefly. Some enzymes cut in the middle of its recogniton sequence, so it produces blunt ends with no overhangs.

Experimental Manual in Medical Biochemistry

## 7.2.2 DNA ligase

In molecular biology, DNA ligase is a particular type of ligase that can link together DNA strands that have double-strand breaks (a break in both complementary strands of DNA). It catalyzes formation of a phosphodiester bond between the 5' phosphate termini of one strand of duplex DNA and the 3' hydroxyl termini of the other. This enzyme will join blunt end and cohesive end termini as well as repair single stranded nicks in duplex DNA, RNA or DNA/RNA hybrids and is extensively used to covalently link fragments of DNA together. Most commonly, it is used with restriction enzymes to insert DNA fragments, often genes, into plasmids vector, which is a fundamental technique in recombinant DNA work. DNA ligases have become an indispensable tool in modern molecular biology research for generating recombinant DNA molecules.

## 7.2.3 Other enzymes for gene cloning

Alkaline phosphotase is used to remove 5'-phosphate group necessary for ligation to avoid the self-ligation of vectors. Klenow fragment is the large fragment of DNA polymerase I, it is used for the conversion of both 5' and 3' protruding termini to blunt ends. RNase H is an endoribonuclease that specifically hydrolyzes the phosphodiester bonds of RNA which is hybridized to DNA. RNase H and reverse transcriptase are both enzymes for making cDNA library.

## 7.3 Vectors for Gene Cloning

There are four major types of cloning vectors: plasmids, bacteriophages and other viruses, cosmids, and some artificial chromosomes. Plasmids are easier to work with; phages and other viruses are more conveniently stored for long periods; larger pieces of DNA can be cloned with cosmids and artificial chromosomes. Besides these major types, there also are vectors designed for a specific function. For example, shuttle vectors are used in transferring genes between different organisms and usually contain different origin for each host. All vectors share several common characteristics. They are typically small, well-characterized molecules of DNA. They contain at least one replication origin and can be replicated within the appropriate host, even when they contain foreign DNA and they always have multiple cloning sits (MCS) allowing the insertion of foreign DNA. Finally, they code for phenotypic trait that can be used to detect their presence.

Expression vectors require not only transcription but also translation of the cloned gene, thus require more components than simpler cloning vectors. In addition to having common vector features such as origin of replication, MCS and selectable marker(s), expression vectors should contain a strong promoter and translational control sequences (Figure 7-2 and Table 7-1). The cloned gene should contain or be engineered to contain an open reading frame (ORF), which starts with a translation initiator ATG and stops with a stop codon. The promoter is positioned immediately upstream of the MCS where the foreign gene is inserted. It drives the transcription of the cloned gene, and often constitutes the major determinant for the degree of expression level obtained. The translational control sequences such as ribosomal-binding site are required for the efficient translation of the transcribed mRNA. The combination of the promoter, cloned gene and translational control sequences is often referred to as an "expression cassette". Moreover, in some vectors, tag sequences are introduced into the expression vector to facilitate the detection and purification of the recombinant proteins (Table 7-2).

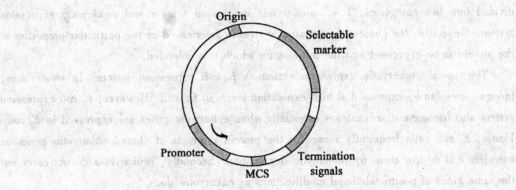

Fig. 7-2 The general structure of expression vectors

Table 7-1      **Features of expression vectors**

| Features | Prokaryotic expression vector | Eukaryotic expression vector |
| --- | --- | --- |
| Selectable marker | Ampicillin resistance<br>Tetracycline resistance<br>Kanamycin resistance | Neomycin resistance<br>Puromycin resistance<br>Blasticidin resistance |
| Promoter | Trp-lac<br>T7<br>Phage T5 | CMV<br>Beta-actin<br>EF-1α |
| Translational control sequences | Shine-Dalgarno sequence | Kozak sequence<br>Polyadenylation signal |

## 7.4 Expression of Foreign Genes and Purification of Recombinant Proteins

### 7.4.1 Expression of foreign genes in host cells

In many cases, gene cloning is employed to express the desired protein for subsequent applications such as functional analysis of novel proteins and therapeutic uses. An expression system mainly consists of an expression vector, its cloned gene, and the host cells which provide the necessary conditions to allow the cloned gene express.

Currently, there are several well-developed expression systems readily available from commercial sources or public repositories. In general, these expression systems are mainly divided into two categories, i. e. prokaryotic expression system and eukaryotic expression system. Generally, the choice of expression system is determined by the particular properties of the protein to be expressed and the purpose for which it is intended.

The typical prokaryotic expression system is *E. coli* expression system. In most cases, foreign genes can be expressed at high expression levels in *E. coli*. However, *E. coli* expression system also has some disadvantages especially when eukaryotic genes are expressed in *E. coli*. Firstly, *E. coli* cells frequently recognize the protein products of cloned eukaryotic genes as outsiders and destroy them by proteolytic degradation. Secondly, prokaryotes do not carry out the same kinds of posttranslational modifications as eukaryotes do.

Eukaryotic systems have some advantages over their prokaryotic counterparts for producing eukaryotic proteins. Firstly, eukaryotic proteins made in eukaryotic cells tend to be folded properly, so they are soluble, rather than aggregated into insoluble inclusion bodies. Secondly and more importantly, eukaryotic proteins made in eukaryotic cells are usually properly modified (phosphorylated, glycosylated, etc.) in a eukaryotic manner, and they almost always accumulate in the correct cellular compartment. However, compared with prokaryotic expression system, eukaryotic expression system are technically difficult, time-consuming, expensive, and relatively slow because eukaryotic host cells are more difficult to manipulate, which have higher demand on media and grow slower than prokaryotic host cells.

### 7.4.2 Purification of recombinant proteins

In general, purification of recombinant protein is relatively easier than normal extraction and purification of proteins from tissues as the recombinant protein is expressed at high level in host cells. The strategy of purification for recombinant protein depends not only on the physicochemical properties of proteins but also the expression manner of the foreign gene (e. g. insoluble inclusion bodies, soluble recombinant proteins, or fusion proteins).

#### 7.4.2.1 Purification of recombinant proteins from insoluble inclusion bodies

High-level expression of many proteins in *E. coli* might result in insoluble inclusion bodies that are failed folding intermediates. However, the formation of inclusion bodies is not necessarily undesirable, and it even has several advantages in some cases. At first, inclusion bodies usually represent the highest yielding fraction of expressed protein. More importantly, inclusion bodies are easy to isolate to high purity by simple centrifugation which will tremendously facilitate the purification of the target protein. In addition, the expressed proteins in inclusion bodies are generally protected from proteolytic breakdown.

If the application is to prepare antibodies, inclusion bodies can be used directly as an antigen without further treatment. However, if the purpose of recombinant expression is to produce functional proteins, the isolated inclusion bodies should be solubilized and refolded by a variety of well-developed methods. Solubilization and refolding methods usually involve the use of chaotropic agents, co-solvents or detergents. For example, once purified, inclusion bodies can be solubilized by denaturation with guanidine hydrochloride or urea.

#### 7.4.2.2 Purification of soluble recombinant proteins

If the cloned gene is expressed as soluble form, it may be excreted or located in the periplasm, in the membrane fraction, or most commonly in the cytoplasm. As for such soluble proteins, the conventional protein purification strategies should be carefully followed. In general, the purification procedures are divided into two main stages: ① based on the localization of the recombinant proteins, different methods such as centrifugation and filtration are used to obtain an extract containing the desired recombinant protein in soluble form; ② multiple conventional purification methods including salt fractionation, chromatography, and electrophoresis may be carried out to get the purified protein. Moreover, for therapeutic proteins, more strict steps must be taken to remove minor contaminants.

#### 7.4.2.3 Purification of fusion proteins

In some cases, the foreign gene is expressed as fusion protein to achieve efficient translation and improved folding, and avoid proteolytic degradation. In this approach, the foreign gene is introduced into an expression vector 5' or 3' to a carrier sequence which codes for a carrier peptide or protein (fusion partner). The fusion partner can be as small as a short peptide such as FLAG and HA, and can also encode an entire protein such as glutathione-S-transferase (GST). Most of the common fusion partners double as affinity tags, and these make the isolation of the expressed protein relatively easy. Proteins can often be purified to >90% by affinity chromatography without knowing their physicochemical characteristics in details. After purification, the fusion partner can be removed by specific proteolytic or chemical cleavage. Detection of fusion proteins is also a simple matter, since antibodies and colorimetric substrates are available for several of the commonly used fusion partners. Table 7-2 summarizes some of the characteristics of the most widely used fusion partners.

Table 7-2  Commonly used fusion partners

| Tag | Tag Size | Purification | Detection | Cleavage |
|---|---|---|---|---|
| FLAG | 1 kU | Anti-FLAG resin | Anti-FLAG antibodies | Enterokinase |
| 6 × His | 1 kU | NTA-agarose | Anti-6 × His antibodies | Enterokinase, Thrombin |
| Myc | 1.2 kU | Anti-Myc antibody resin | Anti-Myc antibodies | Not available |
| Glutathione S-transferase, GST | 26.5 kU | Glutathione sepharose/agarose | Anti-GST antibodies, CDNB substrate | Thrombin, Factor Xa |

Chapter VII  Genetic Engineering

## Experiment 21  DNA Cloning

DNA cloning involves separating a specific gene or DNA segment from its larger chromosome and attaching it to a small molecule of vector DNA. The basic process of DNA cloning includes: isolation of target gene, selection and construction of vectors, digestion and ligation of target DNA and vector, transformation of target gene into receptor cell, screening for recombinant plasmids, and expression of cloned gene, etc.

## Part I  DNA Digestion by Restriction Endonuclease

### Principle

Restriction endonuclease is able to identify specific DNA sequences and cuts the double-stranded DNA within or near identified sites. There are two kinds of cut patterns: sticky ends and blunt ends. According to different purposes, two digesting systems may be used such as small amount of enzyme reaction or large number. The former was always used to identify gene fragment, however, the latter was always used to prepare gene fragment.

### Reagents

1. ddH$_2$O (DNase-free).
2. Restriction endonuclease (10 U/μl).
3. 10 × Reaction buffer (the manufacturers normally provide the optimum buffer).
4. DNA (plasmid DNA or genomic DNA).

### Procedure

1. Adding the following solutions to a new Ep tube (sterile).
   7 μl   ddH$_2$O
   2 μl   10 × Reaction buffer
   1 μl   Restriction endonuclease (10 U)
   10 μl  plasmid DNA (1 μg)
2. Incubate at 37 ℃ for 1 ~ 2 h.
3. The digested results are identified by rapid agarose gel electrophoresis analysis, and then

171

decide whether to terminate the reaction.
4. Reaction can be terminated in three ways.
(1) EDTA can be added into a final concentration of 10 mmol/L, through chelating $Mg^{2+}$ to terminate reaction, or 0.1% SDS is added to denature endonuclease and terminate the reaction.
(2) The digested DNA solution is incubated at 65℃ for 20 min, but some of enzymes can not be completely inactivated by heating.
(3) The digested DNA is extracted by phenol/chloroform, and then precipitated by ethanol. This is the most effective method.

## Notes

1. The $ddH_2O$ is for variable volume. The total volume of 20 μl contains 0.2~1 μg of DNA in a typical reaction. If the digested DNA is more than 1 μg, the total volume can be enlarged according to the ratio of the standard system.
2. A newly sterilized tube must be used every time you take enzyme. The plasmid DNA is finally added to prevent cross-contamination of reagents.
3. Most of the restriction endonucleases are stored in 50% glycerol buffer at -20℃. If glycerol concentration is more than 5% in hydrolysis reaction, the activity of restriction endonucleases will be inhibited. So, it should be accurately diluted 10 times or more.
4. The excessive endonuclease (2~5 times) can shorten time of reaction and obtain complete digesting effect, but it will lead to the decreasing of identification to the specific sequence.
5. In order to save restriction endonucleases, you can appropriately increase enzymatic hydrolysis time, but not too long, otherwise it will produce other enzymatic hydrolysis activities.

## Part II  Sticky end Ligation of Recombinant DNA

### Principle

DNA ligase can join the 3' hydroxyl and adjacent 5' phosphate group of DNA molecules to form phospho-diester bond. In this way, the digested target gene and the vector are connected to produce a new molecule called recombinant DNA. The enzyme used to ligate DNA fragments is $T_4$ DNA ligase, which originates from $T_4$ bacteriophage. This enzyme will ligate DNA fragments having overhanging, sticky ends that are annealed together or fragments with blunt ends. Here using the target gene and the vector containing BamH I and EcoR I sites as an example, the digested target gene and vector can form the same sticky ends. The target gene and the vector are separately digested with *Bam*H I and *Eco*R I. And then recover these two digested DNA fragments by agarose gel electrophoresis. Afterwards, join them together by using $T_4$ DNA ligase. The ligation of the double enzyme digestion is also called directional cloning, because the insertion of the target gene into the vector has only one direction. The basic process is indicated as figure 7-3.

## Reagents

1. ddH$_2$O (DNase-free).
2. 10 × RE Reaction buffer (the manufacturers normally provide the optimum buffer).
3. DNA (plasmid DNA or genomic DNA).
4. *Bam*H I (5 U/μl).
5. *Eco*R I (5 U/μl).
6. $T_4$ DNA ligase (2 U/μl).
7. 10 × ligation buffer (the manufacturers normally provide the optimum buffer).
8. DNA (plasmid DNA or genomic DNA).

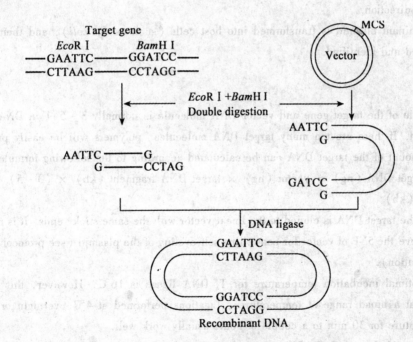

Fig. 7-3  Sticky end ligation

## Procedure

1. Preparing the target gene fragment and vector, taking each 1~3 μg, and then establishing

the following reactions:

| | |
|---|---|
| DNA | 1 ~ 3 μg |
| 10 × buffer | 2 μl |
| BamH I (5 U/μl) | 1 μl |
| EcoR I (5 U/μl) | 1 μl |

ddH$_2$O is added to total volume of 20 μl. The reaction tube is incubated at 37℃ for 12 ~ 16 h or overnight and then the gene fragments and linear vector are recycled by agarose gel electrophoresis.

2. Establishing ligation reaction in 1.5 ml Ep tubes:

| | |
|---|---|
| Target gene fragment | 0.4 μg |
| DNA vector | 0.1 μg |
| 10 × ligation buffer | 2 μl |
| T$_4$ DNA ligase (2 U/μl) | 1 μl |

ddH$_2$O is added to total volume of 20 μl. The reaction tube is incubated at 16℃ for 12 ~ 16 h or overnight. By this method, the target gene is inserted into the vector in the correct direction.

3. Recombinant plasmid is transformed into host cells (such as *E. coli*), and then cloned, screened and identified.

## Notes

1. The ratio of the target gene and vector DNA molecule is normally 3 ~ 5 : 1 in DNA ligation reaction. If there are too many target DNA molecules, polymers will be easily produced. The amount of the target DNA can be calculated according to the following formula:
The target DNA (ng) = vector (ng) × target DNA fragment (kb) × (3 ~ 5) ÷ vector length (kb)
2. When the target DNA is cloned to the linear vector with the same sticky ends, it is necessary to remove the 5'-P of vector for controlling self-cycling of the plasmid (see protocol of blunt-end ligation).
3. The optimal incubation temperature for T$_4$ DNA ligase is 16℃. However, this ligase is active at a broad range of temperatures: Ligations performed at 4℃ overnight or at room temperature for 30 min to a couple of hours usually work well.
4. Both *Bam*H I and *Eco*R I have relatively severe star activity when the ratio of restriction enzyme to DNA is too high. So, you should not use excess amount of enzyme to cut on those vectors, otherwise it might introduce the star activity to the reaction and those enzymes randomly cut somewhere.
5. Heat inactivates the ligase by placing tube in 65℃ water bath for 10 min.

# Part III  Preparation of Competent Cells and Transformation of Plasmid

## Principle

Transformation is to introduce a foreign plasmid into a bacterium and to use this bacterium to amplify the plasmid in order to make large quantities of it or express the target gene. A plasmid is a small circular piece of DNA that may contain ampicillin-, or tetracycline-resistant genes. A plasmid containing resistance to antibiotics is used as a vector. The gene of interest is inserted into the vector plasmid and this newly constructed plasmid is then put into *E. coli* that is sensitive to antibiotic (such as ampicillin). The bacteria are then spread over a plate that contains ampicillin. Finally, the bacteria transformed with plasmids which contain ampicillin-resistant gene can grow in ampicillin plate.

During transformation, recombinant DNA molecules must be introduced into a host to carry out amplification. Since DNA is a very hydrophilic molecule, it won't normally pass through a host cell's membrane or it is lowly taken up by general host cells. Successful transformation is difficult. In order to make bacteria take up the plasmid, they must first be made "competent" to take up DNA. After the bacteria are treated by a high concentration of calcium chloride, its membrane permeability changed so that its uptaking capacity of DNA increases considerably. These cells are called the competent bacteria. The transformed bacteria are cultured in the medium with appropriate antibiotics. The bacteria containing transformant can grow to bacterial colonies.

## Reagents

1. LB liquid medium: pancreatic peptone    10.0 g
    yeast extract    5.0 g
    NaCl    10.0 g

   The substrate is completely dissolved in $ddH_2O$ by using magnetic stirring, and then the pH is adjusted to 7.0 by using 5 mol/L NaOH. The capacity is determined to 1 L, and then autoclaved at 15 Ibf/in$^2$ for 20 min.

2. LB plate: 1.5% agar powder is added into LB liquid medium and autoclaved at 15 Ibf/in$^2$ for 20 min, and cooled to about 60 ℃. Then add ampicillin of 50 ~ 100 μg/ml and pour the solution into culture plate. Put LB plate at 4 ℃ if not used immediately.

3. 0.1 mol/L $CaCl_2$ solution: 1.1 g of anhydrous $CaCl_2$ is dissolved in 60 ml of $ddH_2O$, and then the capacity is determined to 100 ml. The solution is autoclaved, and then stored at 4 ℃.

Experimental Manual in Medical Biochemistry

4. Ampicillin (Amp): 100 mg/ml Amp was stored at -20℃, and diluted to 50 ~ 100 μg/ml for application.
5. Single colony or frozen *E. coli*.

## Procedure

1. Preparation of competent cells
(1) A single colony (e.g. $DH_{5\alpha}$) is transferred to the tube with 3 ml of LB medium, and then shaking at 37℃ overnight. The next day, 1 ml bacterium solution are incubated to a 500 ml flask with 100 ml LB medium, and then vigorously shaking at 37℃ (200 ~ 300 r/min) for 2 ~ 3 h. After $A_{600}$ value reaches 0.3 ~ 0.4, the flask is immediately removed to ice bath for 10 ~ 15 min.
(2) Sterile operation is required from this step. Bacteria are transferred to a 50 ml pre-cooled sterile tube.
(3) Centrifuge (4000g) for 10 min at 4℃, then discard supernatant.
(4) Add pre-cooled 10 ml $CaCl_2$ (0.1 mol/L) to suspend cells. The suspended cells are put in ice bath for 30 min.
(5) Centrifuge (4000g) for 10 min at 4℃, then discard supernatant.
(6) Add pre-cooled 4 ml $CaCl_2$ (0.1 mol/L), and gently resuspend cells.
(7) Immediately dispense every 0.2 ml suspended cells into a 0.5 ml autoclaved Ep tube and stored at 4℃ for 12 ~ 24 h, which can increase competency.
2. To freeze competent cells: Competent cells can be stored at -70℃ for future transformation. Aliquot the competent cells (200 ~ 400 μl), add 30% glycerol, and the competent cells can be stored at -70℃ for several months.
3. Transformation:
(1) Add 100 μl the competent cells into a sterile Ep tube.
(2) Add 50 ~ 100 ng plasmid to the above solution, the volume should not be more than 5% of competent cells. The mixture is gently rotated, and then place on ice for 30 min.
(3) Heat shock at 42℃ for 90 Sec, and then cool the solution about 1 ~ 2 min on ice.
(4) Add 400 μl LB medium without antibiotics into each tube, and shake at 37℃ for 45 ~ 60 min in a shaker (100 ~ 150 r/min). So that bacteria will be reanimated, and express the resistance gene of plasmid.
(5) 200 μl of bacterial solution is spread on LB plate containing Amp of 50 ~ 100 μg/ml. Put plates at 37℃ for 15 ~ 20 min. Then incubate plates upside down at 37℃ for 10 ~ 16 h (not more than 20 h).

## Notes

1. The value of $A_{600}$ should be controlled between 0.3 and 0.4.

2. The used plasmid should be of covalently closed-loop DNA.
3. CaCl$_2$ should be of high quality.
4. Prevent contamination from other bacteria and exogenous DNA.
5. A positive control containing competent cells and plasmid DNA should be established in the experiment to estimate transformation efficiency. A negative control containing only competent cells should be established to eliminate potential contamination. If cloning can grow in a negative control, the possible causes are: ① the competent cells are contaminated by the resistant strains; ② the selective plates failed; ③ the selective plates are contaminated by the resistant strains.
6. No colonies obtained after transformation could be due to inactive competent cells or unsuccessful ligation. Therefore you can check:
(1) Competent cells activity. Perform a test transformation using a known amount of standard supercoiled plasmid DNA as a positive control.
(2) Check the ligase activity. If you have enough ligation products left, analyze linearized vector DNA and ligated product on an agarose gel.
(3) Optimize transformation condition. The added ligation products should be less than 5% of the transformation volume. Excess DNA may inhibit the transformation.

## Part IV  Identification of Recombinant Plasmid

After recombinant plasmids are transformed into *E. coli*, the transformed bacteria need to be screened and identified for selecting recombinant plasmids in which the target gene is inserted according to correct direction. The following are four common methods: ① α-complementary; ② inserting inactivation; ③ in situ hybridization; ④ restriction endonuclease analysis. The first three screening methods only identify if a gene fragment has been inserted in recombinant plasmids, restriction endonuclease analysis can identify inserting direction.

## Protocol I  α-complementary

### Principle

Vector contains the regulatory sequences of β-galactosidase gene (LacZ) and the encoded sequences of 146 amino acids at N-terminus, such as pUC series and pGEM-3Z. A multiple cloning sites are inserted in the encoding region. These vectors can be applied to transform the host cells that can encode C-terminal of β-galactosidase. Neither the encoded fragments by plasmid nor the host cells have enzyme activity, but they can be integrated to recover an enzyme activity. Integrated LacZ can catalyze X-gal to form blue colonies. This phenomenon is called as α-complementary. When the foreign DNA is inserted into multiple cloning sites of plasmid, it will result in inactivation of N-terminal fragment. Therefore, white colonies must contain the recombinant plasmid.

## Reagents

1. X-gal (20 mg/ml): 20 mg X-gal is dissolved in 1 ml dimethyl formamide, and stored at $-20\,^{\circ}\mathrm{C}$ in the dark.
2. IPTG (isopropyl β-D-thiogalactoside) (1 mol/L): 2.38 g IPTG is dissolved in 10 ml $ddH_2O$, filtered by 0.22 μm membrane, and then stored aliquot at $-20\,^{\circ}\mathrm{C}$.

## Procedure

1. Prepare agar plates which contain the corresponding antibiotics (such as Ampicillin).
2. 40 μl X-gal and 4 μl IPTG are added onto a plate surface, and evenly spread by using sterile glass stick. Then incubate at $37\,^{\circ}\mathrm{C}$ for 1 hour to absorb the solution.
3. 200 μl transformed bacteria solution is spread on the plate surface. Put plates at $37\,^{\circ}\mathrm{C}$ for 15~20 min, and then incubate plates upside down at $37\,^{\circ}\mathrm{C}$ for 12~16 h.
4. When stop culturing, incubate the plates at $4\,^{\circ}\mathrm{C}$ for 1~4 h. Blue colonies are negative, while white colonies contain the recombinant plasmids.
5. Pick up white colonies into 5 ml LB medium with corresponding antibiotics such as Amp, and then shake at $37\,^{\circ}\mathrm{C}$ for 8~12 h (200~300 r/min). The plasmids are extracted to further identify by restriction endonuclease analysis.

## Protocol II  Restriction endonuclease analysis

## Procedure

1. Pick up colonies to LB medium with Amp, and then culture them at 37 ℃ overnight.
2. Prepare plasmid DNA by alkaline lysis method (see experiment 16).
3. Restriction endonuclease analysis (see part 1).

# Chapter VII  Genetic Engineering

**Experiment title**
**Date**
**Observations and results**

**Discussion**

1. What is the competent cell?
2. Which characteristic are required for a typical plasmid (vectors) used for molecular biology?
3. How many steps are involved in DNA Cloning?
4. There are four common methods for the identification of recombinant plasmid, which include _____, _____, _____ and _____.

**Teacher's remarks**
**Signature**
**Date**

Experimental Manual in Medical Biochemistry

**Experiment 22**  Expression of Exogenous Gene in *E coli* and Purification of the Expressional Product

## Part I  Detection of Exogenous Gene Expression

### Principle

One of the basic methods used to raise the expression level of foreign gene is to separate the growth of host bacteria and the expression of foreign gene into two phases, which can reduce the burden on host bacteria. The temperature-induce and drug-induced agents are commonly used to improve the expression level of foreign gene. For example, expression vectors with Lac or Tac promoter (such as vector pET-His) can be induced by IPTG. Exogenous protein levels can be detected by SDS-PAGE and Western blotting.

The lysis of bacteria is a key step to extract intracellular products. The application of lysozyme and ultrasonication is popular in laboratory studies. Lysozyme method is to damage the bacterial membrane using lysozyme. Ultrasonication is mechanical crushing, which use ultrasound of more than 15~20 kHz to disrupt cells.

### Reagents

1. LB medium (see experiment 21 Part III).
2. $BL_{21}(DE_3)$ strains of *E. coli* competent cells.
3. Vector pET-His containing the target gene.
4. IPTG stock solution (1 mol/L).
5. Lysis buffer: 50 mmol/L Tris-Cl (pH 8.0)
   1 mmol/L EDTA
   100 mmol/L NaCl
6. PMSF: 50 mmol/L.
7. Lysozyme: 10 mg/ml.
8. Sodium deoxycholate (DOC).
9. DNase I : 1 mg/ml.
10. The reagents for SDS-PAGE.

## Procedure

1. The pET-His containing the target gene was transformed into $BL_{21}(DE_3)$ strains of *E. coli*, and the positive colonies were obtained by screening (see experiment 21, part 3). Plasmid DNA was purified to carry out restriction endonuclease analysis and sequencing.
2. Pick up one positive colony into 3 ~ 5 ml LB/Amp medium, and then shake at 37℃ overnight (200 ~ 300 r/min). Next day, take 0.1 ml bacteria to 10 ml LB/Amp medium, and shake at 37℃ for 2 ~ 3 h (200 ~ 300 r/min). When the value of $A_{600}$ reaches at 0.3 ~ 0.4, add IPTG to a final concentration of 1 mmol/L, and keep on culturing for 3 ~ 5 h. At the same time, the bacteria without IPTG-induced can be as control.
3. Take 10 ml with or without IPTG-induced bacteria respectively, and then centrifuge 5000 r/min at 4℃ for 15 min.
4. Discard supernatant, and resuspend the pellet by lysis buffer (3 ml per gram wet weight of bacteria).
5. Add PMSF 8 μl and lysozyme 80 μl per gram bacteria, and stir using glass stick for 20 min at 37℃. When stirring, add DOC 4 mg per gram bacteria.
6. When the solution becomes viscous, add 20 μl DNase I per gram bacteria. Incubate the solution at room temperature until no longer sticky.
7. (Optional) Bacteria can also be broken further by a sonicator according to the ultrasonic functional parameters. When the supernatant become translucent from turbid, the sonicating finished.
8. Centrifuge 10000g for 15 min, and collect the supernatant and pellet respectively. Take a small volume of the supernatant and pellet, add equal volume of 2 × SDS-PAGE loading buffer to the supernatant and 1 × SDS-PAGE loading buffer to the pellet, pipetting up and down, and heat at 100℃ for 3 ~ 5 min. Then the samples were identified by SDS-PAGE.

## Notes

1. Different expressive plasmids have different promoter, so the inducing methods are not exactly the same. If the vectors have PL promoter, they can be induced by temperature. The bacteria were cultured at 30 ~ 32℃ for a few hours so that $A_{600}$ reached at 0.4 ~ 0.6, then temperature was quickly raised to 42℃, and kept on culturing for 3 ~ 5 h.
2. Ultrasonication should be controlled according to strain type, cell concentration, audiofrequency sonic energy, etc. If the cells were repeatedly frozen and thawed 3 ~ 4 times before ultrasonic treatment, they would be broken more easily.
3. If the target protein is mainly located in the supernatant, it is expressed as soluble form. As for such soluble proteins, the conventional protein purification strategies can be used. If the target protein is mainly located in the pellet, it means the formation of inclusion bodies.

Next are the purification strategies for these insoluble proteins.

## Part II  Isolation, Dissolving and Renaturation of Inclusion Body in *E. coli*

### Step I  Isolation of inclusion body

### Principle

When exogenous proteins are highly expressed in bacteria, they are often gathered to form inclusion bodies in the cytoplasm. After bacteria are lysed, they will be released, and then collected by centrifuging. Inclusion bodies are usually not soluble in water, but can be dissolved by adding protein-denaturing agent such as urea or guanidine hydrochloride. When the expressed proteins contain cysteine, disulfide bonds can be formed in the inclusion bodies. Under such circumstances, the expressed proteins can be completely reduced by denaturant, DTT and β-mercaptoethanol. With the decreasing of denaturant concentration, the expressed proteins can resume their natural configuration, which is called the refolding of proteins with inclusion body. In order to form the correct disulfide bond, the reduced/oxidized glutathione redox buffer is always used to refold. The correct refolding is the most critical and most complex problem in gene engineering.

### Reagents

1. LB medium.
2. $BL_{21}(DE_3)$ strains of *E. coli*.
3. Vector pET-His containing the target gene.
4. 1 mol/L IPTG stock solution.
5. Lysis buffer: 50 mmol/L Tris-Cl (pH 8.0)
   1 mmol/L EDTA
   100 mmol/L NaCl
6. Solution I: 0.5% Triton X-100 and 10 mmol/L EDTA (pH 8.0) were dissolved in the lysis buffer.
7. Urea solution.

### Protocol I  Triton-X 100/EDTA treatment

### Procedure

1. Take 10 ml IPTG-induced bacteria, and then centrifuge 5000 r/min at 4℃ for 15 min.

2. Discard supernatant, and resuspend the pellet by lysis buffer (3 ml per gram wet weight of bacteria).
3. Centrifuge 12000g at 4℃ for 15 min. Discard the supernatant.
4. Resuspend the pellet by adding 9.0 volumes of solution I, incubate at room temperature for 5 min, and then centrifuge 12000g at 4℃ for 15 min.
5. Reserve the supernatant, and resuspend the pellet by 100 μl ddH$_2$O.
6. Take a small volume of the supernatant and resuspended pellet, add equal volume of 2 × SDS-PAGE loading buffer, pipetting up and down, and heat at 100℃ for 3 ~ 5 min. Then the samples were identified by SDS-PAGE.

## Protocol II  Urea treatment

### Procedure

1. Centrifuge cell lysate (see above) 12000g at 4℃ for 15 min. Discard the supernatant.
2. Resuspend the pellet by 1 ml ddH$_2$O per gram becteria. Taking 100 μl ddH$_2$O respectively with 4 Ep tubes, the remaining is reserved.
3. Centrifuge the solution 12000g at 4℃ for 15 min. Discard the supernatant.
4. Resuspend the pellet by 100 μl urea solution [ containing different concentrations of urea (0.5, 1, 2, 5 mol/L) in 0.1 mol/L Tris-Cl (pH 8.5) ].
5. Incubate at room temperature for 5 min, and then centrifuge 12000g at 4℃ for 15 min.
6. Reserve the supernatant, and resuspend the pellet by l00 μl ddH$_2$O.
7. The supernatant and resuspended pellets were detected by SDS-PAGE (see above).

## Step II  Dissolving and renaturation of inclusion body

### Reagents

1. Buffer I: 0.1 mmol/L   PMSF
   8 mol/L   Urea
   10 mmol/L   DTT (dissolved in lysis buffer)
2. Buffer II:
   50 mmol/L KH$_2$PO$_4$ (pH 10.7)
   1 mmol/L EDTA (pH 8.0)
   50 mmol/L NaCl
   1 mmol/L reduced glutathione (GSH)
   1 mmol/L oxidized glutathione (GSSG)

3. 1 mol/L KOH.
4. 1 mol/L HCl.

## Procedure

1. Inclusion bodies (the pellet obtained from step 1) were dissolved by 100 μl buffer I, and put at room temperature for 1 h.
2. Add 9.0 volumes of buffer II, put at room temperature for 30 min, and then adjust pH to 10.7 with KOH.
3. Adjust pH to 8.0 with HCl, and put at room temperature for 30 min at least. Centrifuge 12000g at room temperature for 15 min.
4. Reserve the supernatant, take a small volume, add 2 × SDS-PAGE loading buffer, and then add 1 × SDS-PAGE loading buffer to the pellet. Samples were detected by SDS-PAGE.

## Notes

In order to enhance the efficiency of protein refolding, the following should be noted in dissolving inclusion body: ① maximize the purity of inclusion body; ② fully dissolve inclusion bodies, and reduce proteins completely; ③ if necessary, add reductants; ④ the concentration of denaturants should be reduced little by little, and maintained at the proper concentration to inhibit protein aggregation and wrong intermolecular disulfide bond formation; ⑤ application of redox buffer; ⑥ some factors should be adjusted to the most optimal, such as protein concentration, pH, temperature, the type of salt, ionic strength and so on.

# Part III   Expression and Purification of His-tagged Fusion Protein

## Principle

Affinity chromatography can be performed using a number of different protein tags, such as poly-hisitidine (His) tagged proteins. The histidine tag is very short (6 ~ 10 his residues) and should not alter the conformation of the tagged protein, nor should it be involved in artifactual interactions. The poly-his tag binds to a nickel chelate resin for creation of the column.

The fusion proteins are always the insoluble inclusion bodies (IB), which can prevent from the hydrolysis of protease to the expressed proteins and facilitate to isolate the expressive products using centrifuge. 6 mol/L Guanidine-HCl, 8 mol/L Urea or other strong denaturants can be used to completely solubilize IB. Since under denaturing conditions the His tag is completely exposed, it will facilitate the binding to Ni columns. For most biochemical studies, proteins have to be renatured and refolded, and this can be done in the column itself before elution or after elution.

## Reagents and material

1. Vector pET-His containing the target gene.
2. $BL_{21}(DE_3)$ strains of *E. coli* competent cells.
3. LB medium.
4. 1 mol/L IPTG stock solution.
5. 10 × PBS buffer.
6. Ni-NTA agarose beads.
7. Lysis buffer: 50 mmol/L $Na_2HPO_4$ pH 8.0, 0.3 mol/L NaCl, 1 mmol/L PMSF (or protease inhibitor cocktail for bacterial cells #P-8849 from Sigma) and strong denaturant as 6 mol/L Guanidine-HCl or 6 to 8 mol/L urea.
8. Equilibration buffer: 6 to 8 mol/L urea, 50 mmol/L $Na_2HPO_4$ pH 8.0, 0.5 mol/L NaCl.
9. Washing buffer: 6 to 8 mol/L Urea, 50 mmol/L $Na_2HPO_4$ pH 8.0, 0.5 mol/L NaCl.
10. Elution buffer: 6 to 8 mol/L Urea, 20 mmol/L Tris pH 7.5, 100 mmol/L NaCl, and appropriate imidazole concentrations.
11. The reagents for SDS-PAGE.

## Procedure

1. The vector pET-His containing the target gene was transformed into competent *E. coli*. The transformant was screened in LB plate with Amp (see experiment 21, part 3). Pick the positive colony into 2 ml LB medium with Amp and incubate at 37℃ for 3~5 h.
2. Add 1 mol/L IPTG to a final concentration of 0.1 mmol/L, and then culture at 37℃ for 3~5 h to induce expression of the fusion protein.
3. Centrifuge the bacteria 10000g at room temperature for 5 sec. Remove the supernatant. Resuspend the pellet by precooled PBS of 300 μl.
4. Disrupt the bacteria by ultrasonic, centrifuge 10000g at 4℃ for 5 min, and then transfer the supernatant into a new tube.
5. Resuspend pellet of 10 ml bacterial culture (or 100 ml bacterial culture for very low expression level) in 1 ml lysis buffer.
6. Prepare lysis buffer containing urea 6~8 mol/L or Guanidine-HCl 6 mol/L (try 8 mol/L of urea first, and if protein is soluble titer down in the next experiments till minimal urea is required for protein solubilization).
7. Sonicate on ice 20 sec 3 times (depends of the sonicator). Spin 15 min max speed at 4℃.
8. Transfer supernatant into clean tube: crude extract (keep 40 μl for SDS-PAGE).
9. Equilibrate 50 μl Ni beads with equilibration buffer: Place 50 μl beads (100 μl suspension) of Ni-NTA agarose beads in 1.5 ml plastic tube. Wash 2 times respectively

with 1.5 ml H$_2$O and 1.5 ml equilibration buffer (washing: mix, spin 3500 r/min × 3 min, discharge supernatant).
10. Add the crude extract to the beads and incubate 4℃ for 1 h (swirl).
11. Spin 3500 r/min × 3 min. Discharge unbound material (keep 40 μl for SDS-PAGE).
12. Wash 3 times with 1 ml wash buffer (appropriate urea concentration). Washing: mix, spin 3500 r/min × 3 min, and discharge supernatant (keep 40 μl for SDS-PAGE).
13. Wash 2 times with 1 ml wash buffer + 10 mmol/L imidazole (keep 40 μl for SDS-PAGE).
14. Elute 2 times with 100 μl elution buffer + 100 mmol/L imidazole (keep 40 μl for SDS-PAGE).
15. Final elution: 2 times with 100 μl 250 mmol/L imidazole (keep 40 μl for SDS-PAGE).
16. Run on SDS-PAGE gel 5 μl of crude extract and unbound material, and 13 μl of the wash and elution fractions.

## Notes

1. Do not expose Ni matrices to reducing agents as DTT (you can use β-mercaptoethanol up to 20 mmol/L); chelating agents as EDTA and EGTA; NH$_4^+$ buffers and amino acids as Arg, Glu, Gly or His.
2. Alternate protocol if target protein is not pure enough: Perform parallel purification procedures where you include 10, 20, 30, 40 or 50 mmol/L imidazole in the lysis, binding and washing buffer. Elute directly 3 times with 100 μl elution buffer + 250 mmol/L imidazol. Check eluted proteins on SDS-PAGE. Expect lower yields but higher purification by increasing the imidazol concentration.

## Part IV  Cleavage of His-tagged Proteins with Thrombin Cleavage Sites

### Principle

The fusion gene expressive system has been widely used to produce exogenous proteins in prokaryotic cells. This is completely due to that these systems can generate large amounts of soluble fusion protein. With the universal application of the fusion protein expression, it is particularly important that N-terminal of fusion protein is removed from C-terminal of the target protein. Currently, there are two methods for lysis of the fusion protein in specific site. Chemical method is cheaper and more effective, even can lyse proteins under the denaturing conditions, but it has poor specificity. Enzymatic hydrolysis is more moderate, and highly specific.

Thrombin recognizes the consensus sequence Leu-Val-Pro-Arg-Gly-Ser, cleaving the peptide bond between Arg and Gly. This is utilized in many vector systems which encode such a protease cleavage site allowing removal of an upstream domain. Predominantly the domain to be cleaved is a purification tag such as a 'His-tag'.

## Reagents and material

1. Low-medium salt buffer (20 mmol/L Tris-Cl, 100 mmol/L NaCl at pH 8, no protease inhibitors!).
2. 1% thrombin (purchase from Calbiochem).
3. Benzamidine Sepharose column (purchase from Pharmacia).
4. PMSF (1mmol/L, using a freshly prepared 100 mmol/L ethanolic stock solution).
5. Benzamidine (0.1 ~ 1 mg/ml from a 50 mg/ml aqueous solution).
6. The reagents for SDS-PAGE.

## Procedure

1. After the Ni-column purification it is advisable to exchange (e.g. dialyse) the protein into a low-medium salt buffer. High imidazole (i.e the Ni-NTA column elution buffer) may inhibit the cleavage.
2. Add 1% thrombin preferably at room temperature and shake gently for 1 h. (Use the highest grade human thrombin.)
3. (optional) Monitor reaction the first time by taking aliquots (10 μl) at various time intervals and, via SDS-PAGE look for a gel shift-the protein is 2 kU lower in molecular weight when cleaved so is easily detectable (provided your protein is in the 0-40 kU range). Keep some uncleaved protein as a control and spot a mixture of uncleaved and putatively cleaved in one lane. You may need to allow the gel front to run into the gel buffer solution to get sufficient resolution.
4. Separate the cleaved protein from any uncleaved protein and the His-containing peptide by running the whole mixture through another Ni-NTA column under the same conditions as the first time. The cleaved protein should just flow through and uncleaved protein (which you may wish to recover) will bind to the column and can be eluted with imidazole.
5. Run the collected fraction through a pre-equilibrated (in buffer of choice and with 20 column vol. of equilibration) Benzamidine Sepharose column (1ml of slurry). This should bind residual thrombin preventing secondary activity from this source and some others. Collect the flow through.
6. Add 1 mmol/L PMSF and 0.1 ~ 1 mg/ml benzamidine and any other protease inhibitors you deem necessary.

Experimental Manual in Medical Biochemistry

7. Identified by SDS-PAGE.

**Notes**

A problem with many comercial thrombin sources is secondary protease activity. If this were due to thrombin alone then this would not, in general, be an insurmountable problem, since this can relatively easily be dealt with (with benzamidine and via a benzamidine sepharose column). However human thrombin is one of the most active site-specific proteases. Our experience suggests that such secondary activity arises from other proteases present in the purchased thrombin. Some groups generally further purify purchased thrombin with a Mono-Q ion exchange chromatography step which is an advisable and simple further step.

Chapter Ⅶ  Genetic Engineering

**Experiment title**
**Date**
**Observations and results**

## Discussion

1. Expression vectors require not only _____ but also _____ of the cloned gene, thus require more components than simpler cloning vectors.
2. High levels of expression of recombinant proteins in a bacterial system can lead to the formation of _____.
3. What is the difference between prokaryotic expression system and eukaryotic expression system?
4. What is the advantage if the foreign gene is expressed as fusion protein?

**Teacher's remarks**
**Signature**
**Date**

# English-Chinese Index

| | | | |
|---|---|---|---|
| abscissa | 横坐标 | anionic | 阴离子的 |
| absorbance | 吸收度 | aromatic | 芳香族的 |
| absorption spectrum | 吸收光谱 | atherogenesis | 动脉粥样硬化形成 |
| acetone | 丙酮 | atomic absorption spectroscopy | 原子吸收光谱学 |
| adrenalin | 肾上腺素 | | |
| adsorption chromatography | 吸附层析(法) | atomic emission spectrometry | 原子发射光谱法 |
| affinity chromatography | 亲和层析(法) | autoradiography | 放射自显影法 |
| affinity phase partitioning | 亲和相分离(法) | avidin | 抗生物素蛋白 |
| agar | 琼脂 | bacteriophage T4 | T4 噬菌体 |
| agarose | 琼脂糖 | bed volume | 柱床体积 |
| agarose gel electrophoresis | 琼脂糖凝胶电泳 | bentonite | 皂土 |
| aggregate | 聚集 | benzamidine | 苄脒,苯甲脒 |
| alanine transaminase | 丙氨酸转氨酶 | benzoin acid | 安息香酸 |
| albumin | 白蛋白,清蛋白 | bicinchoninic acid method | 二辛可酸法 |
| alkalescent | 弱碱性的,微碱性的 | biotin | 生物素 |
| alkaline copper sulfate reagent | 碱性硫酸铜试剂 | biuret Method | 双缩脲(反应)法 |
| alkaline phosphatase | 碱性磷酸(酯)酶 | blank solution/reference solution | 空白溶液 |
| ammonium acetate | 醋酸铵,乙酸铵 | blasticidin | 灭菌素,稻瘟素 |
| ammonium persulfate (AP) | 过硫酸铵 | blunt end | 钝性末端,平头末端 |
| ammonium sulfate | 硫酸铵 | borate | 硼酸盐 |
| ampholyte | 两性电解质 | bovine serum albumin | 牛血清白蛋白 |
| ampicillin | 氨苄青霉素 | bromphenol blue (BB) | 溴酚蓝 |
| aneuploid | 异倍体,非整倍体 | butanol | 丁醇 |
| anhydrous | 无水的 | calibration curve | 标准曲线 |
| anion exchange chromatography | 阴离子交换层析 | capillary electrophoresis (CE) | 毛细管电泳 |

| | |
|---|---|
| capillary gel electrophoresis (CGE) | 毛细管凝胶电泳 |
| capillary isoelectric focusing (CIEF) | 毛细管等电聚焦 |
| capillary zone electrophoresis (CZE) | 毛细管区带电泳 |
| carboxymethyl cellulose | 羧甲基纤维素 |
| cation exchange chromatography | 阳离子交换层析 |
| cellulose acetate membrane electrophoresis | 醋酸纤维素薄膜电泳 |
| chaotropic agent | 促溶剂 |
| chelate | 螯合物,螯合的 |
| chloramphenicol | 氯霉素 |
| chloroform | 氯仿 |
| chloroplast | 叶绿体 |
| cholesterol | 胆固醇 |
| chromatofocusing | 色谱聚焦 |
| chromatogram | 层析图谱 |
| chromatography | 层析/色谱 |
| chromophore | 生色团 |
| chylomicron | 乳糜微粒 |
| citrate | 柠檬酸盐 |
| citric acid buffer | 柠檬酸缓冲液 |
| cloning vector | 克隆载体 |
| column chromatography | 柱层析 |
| competent cells | 感受态细胞 |
| concentration | 浓度,浓缩 |
| coomassie brilliant blue | 考马斯亮蓝 |
| coronary heart disease (CHD) | 冠心病 |
| cosmid | 粘粒,粘端质粒 |
| covalent bond | 共价键 |
| crystallization | 结晶 |
| cutoff size | 截阻值(大小) |
| cuvette | 比色杯 |
| dehydration | 脱水作用 |
| dehydrogenase | 脱氢酶 |
| deoxyribonucleotide | 脱氧核糖核酸 |
| desorption | 解析作用 |
| destaining | 脱色 |
| deuterium lamp | 氘灯 |
| development | 展开;显色 |
| dextran | 葡聚糖 |
| diabetes mellitus | 糖尿病 |
| dialysis | 透析 |
| diethanolamine | 二乙醇胺 |
| diethylaminoethyl cellulose | 二乙氨基乙基纤维素 |
| diethylpyrocarbonate (DEPC) | 焦碳酸二乙酯 |
| diffraction grating | 衍射光栅 |
| dimethyl | 二甲基 |
| dimethylbenzene | 二甲苯 |
| dinitrophenyl (DNP)-Protamine | 二硝基苯-鱼精蛋白 |
| disodium phenyl phosphate | 磷酸苯二钠 |
| distribution coefficient | 分配系数 |
| dithiothreitol | 二硫苏糖醇 |
| divinylbenzene | 二乙烯基苯 |
| dyslipoproteinemia | 高脂蛋白血症 |
| E. coli | 大肠杆菌 |
| EDTA | 乙二胺四乙酸,依地酸 |
| electric field strength | 电场强度 |
| electroendosmosis | 电渗 |
| electromagnetic wave | 电磁波 |
| electrophoresis | 电泳 |
| electrophoretic mobility | 电泳迁移率 |
| eluant | 洗脱液 |
| elute | 洗脱 |
| elution volume | 洗脱体积 |
| emission spectra | 发射光谱 |
| end point assay | 终点法 |
| enterokinase | 肠激酶,肠肽酶 |
| enzyme activity | 酶活性(力) |

|  |  |
|---|---|
| enzyme kinetics | 酶动力学 |
| enzyme specific activity | 酶的比活力 |
| enzyme-linked immunosorbent assay (ELISA) | 酶联免疫吸附测定法 |
| equilibrate | 平衡 |
| ethidium bromide | 溴化乙锭 |
| eukaryote | 真核生物 |
| expression vector | 表达载体 |
| extinction | 消光系数 |
| extraction | 抽提 |
| flame photometry | 火焰光度(测定)法 |
| fluorescence spectrometry | 荧光光谱法 |
| fluorimetric assay | 荧光法 |
| formaldehyde | 甲醛 |
| formamide | 甲酰胺 |
| fractionate | 分级,分馏 |
| freezing-thawing | 冻融 |
| gas chromatography | 气相色谱法 |
| gel filtration | 凝胶层析 |
| glucose oxidase | 葡萄糖氧化酶 |
| glutamate-pyruvate transaminase | 谷-丙转氨酶 |
| glutathione S-transferase (GST) | 谷胱甘肽-S-转移酶 |
| glycerol | 甘油,丙三醇 |
| glycerol kinase | 甘油激酶 |
| glycosylate | 使(蛋白质)糖基化 |
| guanidine hydrochloride/guanidine-HCl | 盐酸胍 |
| guanidine isothiocyanate | 异硫氰酸胍 |
| helicase | 解螺旋酶 |
| hemoglobin | 血红蛋白 |
| heparinized | 肝素化 |
| High Performance Liquid Chromatography (HPLC) | 高效液相色谱 |
| homogenate | 匀浆,组织匀浆 |
| homogenization | 匀浆 |
| horseradish peroxidase (HRP) | 辣根过氧化物酶 |
| hydrogen bond | 氢键 |
| hydrogen peroxide | 过氧化氢 |
| hydrophilic | 亲水的 |
| hydrophobic interaction chromatography (HIC) | 疏水作用层析 |
| hypertriglyceridemia | 高甘油三酯血症 |
| imidazole | 咪唑 |
| immunoelectrophoresis (IE) | 免疫电泳 |
| immunoglobulins | 免疫球蛋白 |
| incident light | 入射光 |
| inclusion body | 包涵体 |
| inert | 无活性的,惰性的 |
| infrared spectroscopy | 红外光谱学 |
| inner volume | 内水体积 |
| insulin | 胰岛素 |
| intercept | 截距 |
| ion exchange chromatography | 离子交换层析 |
| ion exchanger | 离子交换剂 |
| IPTG | 异丙基-β-D-硫代半乳糖苷 |
| isoamyl alcohol | 异戊醇 |
| isoelectric focusing (IEF) | 等电聚焦 |
| isoelectric point | 等电点 |
| isopropyl alcohol/Isopropanol | 异丙醇 |
| kanamycin | 卡那霉素 |
| kinetic assay | 动力学分析 |
| lactate dehydrogenase (LDH) | 乳酸脱氢酶 |
| ligand | 配基 |
| ligase | 连接酶 |
| Lineweaver-Burk plot | 双倒数曲线图 |
| lipase | 脂肪酶 |
| liposome | 脂质体,微脂体 |
| liquid chromatography | 液相色谱 |
| liver homogenates | 肝(组织)匀浆 |
| lyophilization | 冻干法 |

| English | Chinese |
|---|---|
| lysogeny | 溶源性 |
| matrix | 基质 |
| matrix-assisted laser desorption-ionization (MALDI) | 基质辅助激光解析电离 |
| mercaptoethanol | 巯基乙醇 |
| michaelis constant | 米氏常数 |
| michaelis-Menten equation | 米曼氏方程式 |
| microplate | 酶标板 |
| minipreparation | 小量制备 |
| miscible | 易混合的 |
| mitochondria | 线粒体 |
| mobile phase | 流动相 |
| molar extinction coefficient | 摩尔消光系数 |
| monochromatic light | 单色光 |
| monochromator | 单色器 |
| n-butanol | 正丁醇 |
| neomycin | 新霉素 |
| neurotoxin | 神经毒素 |
| nitroblue tetrazolium (NBT) | 四唑氮蓝 |
| Northern blotting | RNA 印迹技术 |
| oligonucleotide | 寡聚核苷酸 |
| ordinate | 纵坐标 |
| organelle | 细胞器 |
| organic solvent | 有机溶剂 |
| ortho-Toluidine | 邻甲苯胺 |
| pancreatic | 抑酞酶;胰脏的 |
| pancreatic peptone | 胰蛋白胨 |
| paper chromatography | 纸层析 |
| partition chromatography | 分配层析 |
| peroxidase | 过氧化物酶 |
| phage | 噬菌体 |
| phenazine methosulfate | 吩嗪硫酸甲酯 |
| phenol reagent | 酚试剂 |
| phenol | 苯酚,石炭酸 |
| phosphate transacetylase | 磷酸乙酰基转移酶 |
| phosphodiester bond | 磷酸二酯键 |
| photomultiplier tube | 光(电)倍增管 |
| pipette dropper | 移液滴管 |
| plasmid | 质粒 |
| polyacrylamide gel electrophoresis (PAGE) | 聚丙烯酰胺凝胶电泳 |
| polyadenylation | 多聚腺苷酸化 |
| Polyethylene glycol (PEG) | 聚乙二醇 |
| polymers | 聚合物 |
| polypeptide | 多肽 |
| polysaccharide | 多糖 |
| polystyrene | 聚苯乙烯 |
| polyvinyl pyrrolidone | 聚乙烯吡咯酮 |
| pore | 细孔 |
| precipitate | 沉淀 |
| precipitation | 沉淀作用 |
| precipitin arc | 沉淀弧 |
| prism | 棱镜 |
| prokaryote | 原核生物 |
| protease | 蛋白酶 |
| pulsed-field gel electrophoresis (PFGE) | 脉冲场凝胶电泳 |
| puncture | 刺孔 |
| purine | 嘌呤 |
| puromycin | 嘌呤霉素,博罗霉素 |
| pyrimidine | 嘧啶 |
| qualitative analysis | 定性分析 |
| quantitative analysis | 定量分析 |
| quinone imine | 醌亚胺 |
| radiometric assay | 同位素分析 |
| reaction velocity | 反应速度 |
| relative mobility | 相对迁移率 |
| resin | 树脂 |
| restriction endonuclease | 限制性核酸内切酶 |
| Restriction Fragment Length Polymorphisms (RFLP) | 限制性片段长度多态性 |
| rotary evaporator | 旋转蒸发器 |
| salting out | 盐析 |

|  |  |
|---|---|
| scattering spectra | 散射光谱 |
| sedimentation coefficient (S) | 沉积系数 |
| semipermeable | 半透性的 |
| sephadex | 葡聚糖凝胶 |
| silver staining | 银染法 |
| slit system | 光栅系统 |
| sodium deoxycholate | 脱氧胆酸钠 |
| sodium dodecyl sulfate | 十二烷基磺酸钠 |
| sodium salicylate | 水杨酸钠 |
| solute | 溶质 |
| solvent | 溶剂 |
| sonicator | 超声破碎仪 |
| Southern blotting | DNA 印迹技术 |
| spectrometry | 光谱测定法 |
| spectrophotometer | 分光光度计 |
| spectrophotometry | 分光光度法 |
| stacking force of the base | 碱基堆积力 |
| stacking gel | 浓缩胶 |
| staining | 染色 |
| standard addition method | 标准管法 |
| standard solution | 标准溶液 |
| stationary phase | 固定相 |
| sticky end | 粘性末端 |
| stock solution | 贮存液 |
| substrate | 底物 |
| Sudan black B | 苏丹黑 B |
| sulfonic acid cation exchange resin | 磺酸阳离子交换树脂 |
| supernatant | 上清 |
| tandem mass spectrometry | 串联质谱法 |
| test solution / determined solution | 测定溶液 |
| tetracyline | 四环素 |
| tetramethylethylenediamine(TEMED) | 四甲基乙二胺 |
| thin film chromatography | 薄膜层析 |
| thin layer chromatography | 薄层层析 |
| thrombin | 凝血酶 |
| titanium trichloride-ninhydrin solution | 三氯化钛-水合茚三酮溶液 |
| transmittance | 透射率 |
| triglyceride | 甘油三酯 |
| tungsten lamp | 钨丝灯 |
| two-dimensional electrophoresis (2DE) | 双向电泳 |
| ultrafiltration | 超滤 |
| ultraviolet | 紫外线 |
| uricase | 尿酸酶 |
| vacuum phototube | 真空光电管 |
| vertical | 垂直的 |
| visible | 可见的 |
| void volume | 外水体积 |
| wavelength | 波长 |
| western blotting | 免疫印迹 |
| X-gal | 5-溴-4-氯-3-吲哚-β-D-半乳糖苷 |
| xylene | 二甲苯 |
| yeast | 酵母 |
| zymogram | 酶谱 |
| α-ketoglutaric acid | α-酮戊二酸 |
| β-galactosidase | β-半乳糖苷酶 |
| β-globulin | β-球蛋白 |
| β-Mercaptoethanol | β-巯基乙醇 |
| $\lambda_{max}$ | 最大吸收波长 |
| 1,4-dithiothreitol (DTT) | 二硫苏糖醇 |
| 2,4-dinitro-phenylhydrazine | 2,4-二硝基苯肼 |
| 4-aminoantipyrene | 4-氨基安替比林 |
| 8-hydroxyquinoline | 8-羟基喹啉 |